# City with a Hidden Past

**Fumihiko Maki**

**Yukitoshi Wakatsuki**
**Hidetoshi Ohno**
**Tokihiko Takatani**
**Naomi Pollock**

translated by
**Hiroshi Watanabe**

Kajima Institute Publishing Co., Ltd

City with a Hidden Past

Copyright © 2018 by Fumihiko Maki, Yukitoshi Wakatsuki, Hidetoshi Ohno, Tokihiko Takatani, Naomi Pollock.

All rights reserved. No part of this book may be reproduced in any form
by any electronic or mechanical means without permission in writing from the publisher.

Published by Kajima Institute Publishing Co., Ltd, 2018, Tokyo, Japan.
Translated by Hiroshi Watanabe. Designed by Tatsuki Takaki.

Printed and bound in Japan.
ISBN978-4-306-04661-0 C3052

Contents

| | PREFACE TO THE ENGLISH-LANGUAGE EDITION | | 5 |
|---|---|---|---|
| INTRODUCTION | TOKYO: A SPRAWLING CITY OF SMALL WONDERS | *Naomi Pollock* | 7 |
| CHAPTER 1 | OBSERVING THE CITY | *Fumihiko Maki* | 15 |
| CHAPTER 2 | THE UNDERLYING STRUCTURE OF STREETS | *Tokihiko Takatani* | 39 |
| CHAPTER 3 | MICROTOPOGRAPHY AND PLACENESS | *Yukitoshi Wakatsuki* | 69 |
| CHAPTER 4 | THE EXTERNAL LAYERS OF STREETS | *Hidetoshi Ohno* | 103 |
| CHAPTER 5 | THE JAPANESE CITY AND INNER SPACE | *Fumihiko Maki* | 153 |

# PREFACE TO THE ENGLISH-LANGUAGE EDITION

We are delighted that Kajima Institute Publishing Co., Ltd has decided to publish *City with a Hidden Past,* the English-language edition of *Miegakure suru toshi*. We welcome it for a number of reasons. First, there is the fact that, though comparisons are frequently drawn between Tokyo and metropolises such as New York, London, Beijing and Paris with respect to cultural, political and economic matters, relatively few people are familiar with the qualities that make Tokyo interesting as a city, especially the morphological aspects of its historical development and present condition. We have hopes therefore that this book will help many more people understand Tokyo.

The original work was based on a study commissioned by the Toray Science Foundation on "The Desirable Living Environment" and undertaken by a team composed of Fumihiko Maki, Morikazu Shibuya, Yukitoshi Wakatsuki, Hidetoshi Ohno and Tokihiko Takatani (all members at the time of Maki and Associates), as well as Masaharu Rokushika. The parts of the study having to do with Japanese urban spaces were then published in 1980 as *Miegakure suru toshi*, and the book has subsequently gone through twenty-one printings and attracted many readers.

On this occasion, we asked our friend Naomi Pollock to draw on her many years of experience living in Tokyo to present her view of what makes the city unique; her discussion gives the book an added dimension. We also asked Hiroshi Watanabe, our mutual friend, to render the text into English, and he has produced a clear and thoughtful translation.

More than thirty years have passed since the Japanese-language edition was first published, and Tokyo has continued to undergo rapid changes. Not many places still appear as they did then. However, historical awareness of the city's order has not disappeared. In particular, the concern for surface layers, microtopographies and small spaces in the city—with which this book deals—can often be glimpsed in the aesthetic consciousness that is distinctive to Tokyo, even when the places themselves have been physically transformed.

Also important is the fact that in this age of globalization, many more of us are acquiring the perspective to observe the metropolises of the world in an impartial way. Today, enormous new cities with populations in the millions or tens of millions are emerging, mainly in developing countries. Twenty years ago, who could have imagined that Shenzhen, China, would develop into a city of almost 12 million? One also finds in that country a city such as Chongqing where medium-rise housing is introduced directly under a high-speed railway and a monorail passes between high-rise apartment buildings. Such images reveal a fundamentally different attitude toward public domains in cities. They are sharp reminders that, as the result of globalization, old ideas about the city based mainly on traditional Western society no longer apply. If cities and buildings are expressions on the part of those living in them of a desire for spatial order, then, as the different forms that order takes proliferate, the actions of those who create those forms may be best understood through comparative cultural anthropology. This book may provide an opportunity for people throughout the world to understand the nature of Tokyo's order from a macro point of view, recognize its special characteristics and understand its past and present. We would like to express our deep gratitude to the Kajima Institute Publishing Co., Ltd, which gave us its full support in bringing out this English translation, especially to Kōji Aikawa for playing a central role in the undertaking.

The Authors<br>Autumn, 2017

INTRODUCTION

## TOKYO:
## A SPRAWLING CITY OF SMALL WONDERS

*Naomi Pollock*

As an architect-turned-journalist, I am often asked how I can write about beautiful design in a place as ugly as Tokyo. Indeed, the city does not always make a good first impression on visitors from abroad. What they see is a chaotic confluence of concrete surfaces, irregularly shaped buildings, crowded streets, above-ground utility poles and garish signage. And on top of that, a web of elevated expressways casts unrelenting shadows on many of the city's major thoroughfares.

This presentation contrasts sharply with the tourism industry's clichéd portrait of Japan: wooden buildings in idyllic settings. It also differs from the familiar conditions back home in Europe or the United States. Conspicuously absent are the open squares and gracious boulevards that govern many a Western city. And instead of a carefully considered waterfront, Tokyo basically turns its back on miles of shoreline and riverside. Not to mention the canals that once riddled the city. Converted to paved traffic conduits, they are now traversed by a steady stream of cars.

Against this cacophonous backdrop, few individual buildings stand out. And those that manage to do so often have little regard for material consistency or geometric compatibility with their neighbors. Here commercial buildings vie aggressively for attention while many residential ones either disregard their surroundings completely or suffer from nondescript monotony. If one travels to Tokyo expecting to find the architectural harmony of New York's Fifth Avenue or the elegant unity of Boston's Beacon Hill, disappointment will surely abound. But if the eye can be retrained to focus on the small—a narrow walkway, a pocket park, or even a single window—Tokyo becomes one of the most magical places on earth.

It is not that big, urban gestures do not exist in Tokyo. They certainly do. Underlying today's dense development is a patchwork of localized grids and other organizational devices, many remnants of the Edo period. Despite the city's complete devastation and reconstruction necessitated by catastrophic manmade and natural disasters, Tokyo's intricate web of streets—broad avenues spanned by narrow streets connected by pedestrian pathways—still reflects the historic social hierarchy and land divisions. In some parts of the city, such as Marunouchi, streets form a recognizable, rectilinear grid. Elsewhere, they assume a more organic configuration that accommodates to the Kanto Plain's topographical features.

Unlike in American cities, where the street grid defines the coordinates of the address system, a sequence of concentric circles is a more apt descriptor when it comes to locating a particular building in Tokyo. Experientially, many residential neighborhoods are organized around transportation hubs, ringed by shops, surrounded by apartments and houses. Since relatively few

*Left:* Street in Harajuku, Tokyo.
*Right:* Street in Kyojima, Tokyo.

Tokyo streets have names, local landmarks and visual clues, such as stoplights, schoolyards and convenience stores, are the keys to getting from point A to point B. By navigating Tokyo through this lens of wayfinding, its microscopic beauty begins to come into clear focus.

In exchange for careful study of the surroundings, one is apt to discover a city where craftsmanship still abounds, greenery crops up in unexpected places and people take meticulous care of their environment. One need not be an architect to delight in a well-loved wooden house nestled between a gas station and a neon-clad convenience store, catch a glimpse of an open entry foyer or even notice a beautiful detail where wood expertly joins concrete. If you look—really look—there are small treasures to behold throughout this city.

In part this is due to Tokyo's diminutive scale. Though geographically sprawling, the city is an agglomeration of tiny pieces. Instead of building upwards, Tokyo has a habit of expanding outwards within the boundaries delineated by mountains on three sides and the ocean on the fourth. While outcroppings of tall buildings keep popping up in commercial districts, Tokyo remains a relatively low-scale city.

Instead of making increasingly bigger buildings, available property is often divided into smaller and smaller parcels. This is especially prevalent in residential districts, resulting in more and more houses on less and less land. While large apartment complexes are "home" to many, Tokyo presents as a city of individual houses, most not much bigger than 100 square meters and many considerably smaller.

To some extent this habit of consecutive subdivision reflects Japan's real estate economics. In Tokyo, more often than not, the land is many times more valuable than the house itself, creating an incentive to slice and sell. But it also speaks to the national desire for a plot of one's own. Even in the middle of Tokyo, plenty of residents hanker after a direct connection to the ground, a

Street in Tsukishima, Tokyo.

desire rooted in the primacy that nature holds in Japan.

This may also explain the national penchant for things that grow. Many an open gate reveals a lovely garden, but those with less or no land make do with potted plants. Adding a dash of seasonal color, lovingly nurtured flowers dot stoops, porches and balconies citywide. In contrast to the United States, where bigger is almost always better, in Japan a single blossom is often deemed more precious than an entire flowerbed.

But the appreciation of nature is not limited to the individual homeowner. It is manifest at the urban scale as well. Even in this city of concrete, greenery and trees are surprisingly plentiful and usually well tended. While some trees enjoy special preservation status, others fall prey to development. Yet new construction must comply with the city's stringent code requirements. While the inclusion of greenery may be regarded as beneficial, the protection of daylight access is immutable. In addition to codes controlling the size and density of new construction, the so-called "sunshine laws" ensure that the sun's rays reach street level for a portion of the day, even in winter when the sun is at its lowest. These laws account for the strict height limitations and the bizarrely triangulated rooftops that result from maximizing the legal building envelope. Though the architectural implications of these laws do not always yield the most elegant solutions, the omnipresence of sunshine is a delight.

Perhaps the most immediate evidence of nature's high stature is the prevalence of public space citywide. Running the gamut from large, municipal gardens to neighborhood parks and tiny playgrounds, a very wide and plentiful range of green spaces is available to Tokyo citizens who make good use of this urban asset. In the early morning the city's parks may belong to the elderly participating in group exercise programs but by midday stroller-pushing mothers and their small tots take over, followed by students of all ages once school lets out and, throughout the day, dogs and their owners. In a city

Two views of a pocket park in Tokyo.

where so many people do not have the space for gardens of their own, parks function as everyone's backyard.

When it comes to livability in a city as developed as Tokyo, gaps and glitches in the urban fabric are in many ways as important as the buildings themselves. In addition to the planned park system, the city is laced with an array of void spaces of various types. Evoking religious structures located out in the country, shrines and temples of all different sizes are embedded in the city, most approached via tree-lined walks reminiscent of the ritual, wooded path. A vast tree-laden precinct, Meiji Shrine is a remarkable oasis just a stone's throw from Harajuku, one of Tokyo's liveliest neighborhoods. Elsewhere, tiny *torii* gates separate the secular from the spiritual, providing a respite from the hustle and bustle.

Yet in most instances, openings in the urban fabric result from more practical considerations. For fire safety, party walls are largely prohibited and physical separations, some just wide enough for a person to pass, are required between buildings. This system sets up a marvelous, syncopated cadence of

Meiji Shrine, Tokyo.

solid and void. Though they often double as storage, these slivers of space are much-needed pauses that provide metaphorical breathing room. In perimeter neighborhoods, agricultural plots, where local citizens cultivate vegetables in neat rows, occupy open parcels and coexist amiably with the surrounding structures. In more central neighborhoods, leveled land appears practically overnight when outdated buildings are demolished. Posing as parking lots, most are construction sites waiting to happen.

Unsurprisingly, the ephemerality of buildings lends itself to an on-going, organic form of redevelopment. Here houses rarely last more than thirty years. And buildings of all types are torn down without a shred of sentimentality. Though it comes at the cost of the city's architectural heritage, the bit-by-bit replacement of the old with the new is invigorating. Unlike in many municipalities in the United States, continual, small-scale reinvention means that Tokyo is not at risk of becoming a deserted donut city once the workday is over.

In addition to the reconstruction of individual buildings, several large, mixed-use developments have sprung up around the city. Working in tandem with the existing texture of the city, Marunouchi is being redeveloped building by building. Like teeth, single structures are pulled and replaced one at a time, retaining the neighborhood's atmosphere but updating it at the same time. Elsewhere, other large undertakings have followed different growth models. While Tokyo Midtown occupies a large parcel acquired in one fell swoop from its previous owner, the Japanese Government, other projects, such as Roppongi Hills, necessitated the purchase of many small properties belonging to numerous owners. This lengthy process took years to complete even before ground could be broken. Entirely transforming the site, Roppongi Hills turned a quiet residential enclave into a commercial hub defined by an

Two views of Marunouchi, Tokyo.

office tower, high-rise apartment buildings, outdoor plazas for the public and a multi-level, indoor shopping mall.

To a certain extent, big developments are inevitable in Tokyo. Land in the center of the city remains a much sought-after commodity, and building tall is one way to maximize its occupancy. But this comes at a cost. Tokyo's human scale is one of its most unique traits and finest assets. As a writer and a designer, I can appreciate, and even enjoy, the new, super-sized construction. But Tokyo's treasure trove is hidden in and among its low buildings and circuitous streets.

CHAPTER 1

# OBSERVING THE CITY

*Fumihiko Maki*

What does it mean to understand the form of a city? The answer to this seemingly plain question is by no means simple. To understand the history, social organization and economic system of a city, for example, is not to know and remember every phenomenon that has occurred there. Instead, it is first necessary to discover what the most important principles underlying those phenomena are. Only then will we understand the significance of and the interrelationships between phenomena.

The same can be said of the form of a city. No one is fully cognizant of all the diverse elements from which a large city is composed, nor is there need for such knowledge. Therefore, we grasp first the whole and then the parts by means of shapes or images that have been sorted out in some way. For that, there are maps, photographs, and from earlier periods, picture scrolls and wood-block prints. In recent times, reconstruction models have helped further our understanding, but they are still far from sufficient. We are relatively well stocked in material concerning certain aspects of the Japanese city such as the look of and the nature of life in Kyoto in the Heian period, the castle towns that developed in the Sengoku (Warring States) period and areas where artisans and tradesmen lived in Edo, but such evidence is, from an overall point of view, quite fragmentary at best.

By contrast, we can examine with our own eyes the city we live in at present. Ample material—perhaps even an excess of material—exists on the subject. The problem first of all is to know what to look at, and unless we have a point of view, we cannot get a true understanding of form or morphology. Anyone can pass judgment or have an opinion on the beauty or ugliness, the appeal or absence thereof, of a city as a whole or its parts. However, that is not to truly understand the city. As with economics or history, it is first necessary to know the principles behind the generation of diverse forms.

If we look at the history of painting, there is a clear difference of vision— that is, of subject matter—between artists of the past who depicted heroes, wars and other dramas in the world of humankind and modern artists from the Impressionists and Cubists to the practitioners of Pop Art. Needless to say, a kind of vision or viewpoint is similarly required to understand the city.

Urban morphology is perceived as an accumulation of elements, such as the mountains that surround the city, the rivers that flow through the town, the arrangement of streets and blocks, and the fenestrations and ornamentations of buildings. People can move freely amidst these elements and receive and record impressions of individual or continuous changes in form. However, as I stated at the outset, a viewpoint that has been adjusted in some way is necessary, given the breadth of things in the city we may see and from which we may receive impressions.

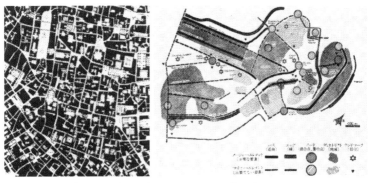

Fig. 1-1 *Left:* Map of the center of Rome by Giovanni Battista Nolli, 1748.
Fig. 1-2 *Right:* Perceptual form of Boston. From Kevin Lynch, *The Image of the City*.

People have therefore dealt for various reasons and by various means with the fact that the whole is difficult to see even though individual elements may be easily perceived. The Nolli Map (Fig. 1-1), which is well known among architects and frequently referenced in this book, showed seventeenth-century Rome from a fresh perspective based on a figure-ground relationship. That is, it showed quite clearly with what morphological intentions that city was created by rendering buildings in dark poché and leaving exterior spaces such as roads and squares white. Furthermore, it revealed the composition of public domains and private domains in Rome: the interiors of public buildings were left white just as outdoor squares and street spaces were.

In the 1950s, the American urbanist Kevin Lynch, by monitoring and analyzing the way people observed cities, established that certain formal elements in cities left the greatest impression on people. He showed that an image map (Fig. 1-2) could be drawn of any city in the world with five elements: paths, edges, districts, nodes and landmarks. His method based on image was of great importance in demonstrating that it was possible to transcend cultural differences in expressing the way we look at a city.

Many architects and urbanists have subsequently developed various methods and proposals for understanding urban morphology. Although these studies have undeniably contributed in some measure to our understanding of the external characteristics of forms and impressions made by forms, they have failed to delve deeper and cast light on the intentions and meanings behind urban forms. That is, we cannot achieve true understanding until we comprehend at the same time what each urban form means within the cultural context of urban society.

In particular, indicating figure-ground relationships as Nolli did is by no means a suitable method for uncovering the character of cities that owe their development to a mindset or ideas entirely different from those on which Western cities are based, especially when as in Japan those cities are characterized by "gaps," as will be explained later. The same can be said of Lynch's image map.[1]

The most important aim of research into urban morphology is to reveal, through an analysis of patterns and shapes, why and by what means a certain form or structure came into being. Clearly in many cases the resulting urban form or structure is not a pure expression of intention but the imperfect consequence of various factors and happenstance. That is, the intentions and calculations of many different parties are at work in a city. However, such imperfections always characterize the processes by which a city is created and are not undesirable. Rather, they render a city more interesting, endowing it with unexpectedness, newness, ambiguity and irony. The more such divergences exist, the higher the context of the city as a whole, as Edward T. Hall explains in *Beyond Culture*, published in 1976.[2]

One way to seek a better understanding of urban morphology is to discover what might be called the deep structure existing behind the imperfect expressions constituting the surface of urban form. We gradually gain an understanding of structure by clarifying the principles unique to a city or the community from which that city developed, or the relationship between those principles. The principles may be morphological or spatial, but if we can discover the reasons behind those principles—that is, if we can discover in many cases the cultural meanings of those principles—we can elucidate the relationship or structure between elements.

For example, the grid pattern—frequently discussed in the following chapters on street patterns, adaptation to microtopography and *oku*—clearly has a different meaning and structure for each community. An arrangement in which urban territory is demarcated and city blocks are created on the basis of two orthogonal axes can be said to be among the most universal and commonsensical patterns made by human beings, one that transcends regional and cultural differences. However, the grid pattern of Miletus, said to have been created by Hippodamus was abstract and did not acknowledge topographical changes in the surrounding area. It is said to have expressed the equality of citizens in the Greek city-state. On the other hand, the grid pattern of Manhattan was created from a thoroughly practical perspective. The grid pattern there is in the nature of a guideline permitting buildings independence and freedom within the limits it sets. Manhattan might be likened to an orderly set of cages for wild beasts, that is, skyscrapers (Fig. 1-3).

Though likewise American, the nine-square grid organized around a

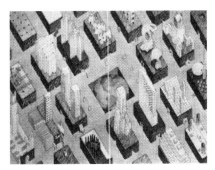

Fig. 1-3 The City of Captive Glove. From Rem Koolhaas, *Delirious New York*.

commons found in early New England towns was expressive of a strong will to encircle a center. That is, though also a grid pattern, it reflected, as Gustav Wolf points out, group memory of a time when villages were formed around a green for livestock.

Aside from cities in the ancient period such as Kyoto and Nara that were based on the walled cities of China, many castle towns including Edo had grid patterns. As a rule, a grid pattern was used to indicate a residential area of people of the same social status. For example, estates of samurai of the *hatamoto* rank, estates of lower-ranked samurai such as *yoriki* and *dōshin*, and districts of *chōnin* were all defined by grid patterns, but the location within the city and the size of a particular grid reflected the social status of the residents. In particular, the blocks assigned to untouchables (*burakumin*) found in old castle towns, though also based on a grid pattern, were often far smaller in dimension compared with other blocks, confirming that there was indeed an intent to form a separate territory for each social rank. Furthermore, as is explained in Chapter 3, the orientation and size of the grid pattern changed freely depending on factors such as natural topography and landscape. And as both the history of Edo (the old name for what is now Tokyo) and the historical changes in the original grid layout of Kyoto show, the grid pattern in Japan was in many cases subdivided in various ways until it gradually became nothing more than a guideline.

Understanding the grid pattern is thus the first step in reading the structure behind it. We are apt to pigeonhole our impressions of a city, for example, simply equating a grid pattern with orderliness. The question here is not one of opposition to or approval of an orderly townscape; everyone is entitled to his or her opinion on that point, just as a community is entitled to its view. The danger is that in understanding the grid pattern simply by its attribute of orderliness, we fail to delve deeper and see the urban context of that pattern.

We are not necessarily seeing the structure behind urban form, but neither are we entirely blind to it. The structure remains elusive.

This situation is similar to the relationship between the initial sketch drawn by an architect designing a building and the building as ultimately constructed. The sketch made in the first stage of design before many conditions have been presented is in many cases the most direct expression of the architect's intention. However, during the design process, various conditions such as changes in the client's program and budgetary and regulatory constraints gradually transform the first sketch until in the end it is seemingly something quite different. However, when we compare the initial sketch with the final design, the original intention or the character, philosophy or mode of operation of the architect can often be glimpsed in the scheme as built. I have already stated that urban form, unlike architecture, is the result of the manipulations of a collective will and not the will of an individual. Yet to the extent that urban form is the result of the manipulations of a collective will, it reveals in diverse forms the decisions of the group belonging to a particular culture. Urban structure can be regarded as a rough plan that is elaborated by the collective will within a culture unique to a community.

To that extent, rough plans are more traditional and custom-bound than individual buildings; the deep structure itself is not subject to endless change or sudden mutation. That is, though there may be radical transformations on the surface, there is stubborn resistance to change on a deeper level. Seen in this way, reading the relatively stable "underdrawing" beneath urban forms can be said to be for now the first step in understanding the city.

## THE FORMALIZATION OF NATURE

What are the characteristics of the urban structure we have lived in from early modern times to the present day? Unlike cities in other societies such as the West and China and in desert regions, which are materializations of urban images based on abstract concepts, Japanese cities have been created through a process in which nature or formalized nature has been actively used on diverse levels. I would like to emphasize that far from being the product of a passive, accidental process, they were the result of a quite deliberate process of urban development in a unique context.

Formalization of nature is not an unusual phenomenon in itself; it is to be found in various guises in every culture since ancient times. However, in the development of cities in Japan, formalized nature serves—as do the plants on the external surfaces of cities discussed in Chapter 4—as an important, indispensable element in the creation of spatial order from the scale of the individual building to the urban scale. It has even been regarded at times as belonging to a higher order than the man-made order. It should be noted, moreover, that formalization of nature was quite unique in Japan in that in certain fields it achieved a high degree of refinement and followed patterns lasting over centuries.

In Western urban and architectural culture, nature has been regarded as being in opposition to, and not infrequently of a lower order than, what was man-made. Even Rococo and Art Nouveau, which adopted natural forms as motifs, were reactions to, or refutations of, a constant emphasis on the abstract and the geometrical. Such movements may have briefly made their mark on the city but did not by any means have the power to change the appearance of the city as a whole. By comparison, nature has remained important to the city for virtually the entire course of urban history in Japan.

Vernacular architecture of any community is based on various principles and conventions that have been accumulated, modified and sorted out over time. Many of these are deeply rooted in climate, religion and social customs and suggest the meaning behind forms.[3] That is, formalized nature has shaped in various ways many principles of architecture as well as urban development in Japan.

Building types such as townhouses (*machiya*), samurai estates (*buke-yashiki*) and farmhouses (*nōka*) and rules such as *kiwari*, which govern dimensions of building elements have developed over time. The idea of *kasō* or geomancy is also noteworthy. Known in China as feng shui, it has prescribed the relationship between topography and architecture since ancient times. Feng shui was originally a system of principles determining the location and orientation of graves but its scope of concern eventually came to include buildings and even city planning. The ideas of feng shui influenced palatial architecture in Japan and eventually spread to the rest of society, reaching its full development in Japan in the Edo period. As is explained by the architect Kiyoshi Seike in his bestseller *Kasō no kagaku* (The Science of Geomancy),[4] geomancy includes many principles that have rational bases even though we are apt to dismiss it today as superstition.

Why did such rational elements combine with irrational elements and develop into geomancy in China and Japan? An interest in human fate that was in ancient times the province of religion and divination was undoubtedly closely tied to the irrational elements of geomancy. Geomancy began as an attempt to interpret diverse phenomena in the natural world through the yin yang theory and the five elements or phases. On the one hand, it created through the elucidation of phenomena a foundation for the study of the environment, and on the other, it developed through symbolism into a view of religion and civilization. As Seike explains, *kasō*, stripped of its ties to divination, is essentially a manual for building a house, one that shows how to select a site appropriate to the climate, how to design the area around a house and how to determine the layout, structure and materials of a house. In particular, its rules on the selection of a site, the design of the area surrounding a house, and the relationship with the natural environment are as much about aspects of the city as they are about aspects of the house. That is because, when the man-made environment comes into contact with nature, the stance taken ought to be basically the same whether that environment is a single building, a town or a city. We can easily imagine therefore that in the development of cities in Japan, rules of geomancy concerning what was deemed felicitous or infelicitous in the lay of a land and the configuration of a city played an important role in determining the siting and scale of various urban functions as well as the urban layout. For example, the planners of Edo invariably selected places on solid ground for important facilities such as Edo Castle and the Higashi Honganji temple even though they did not have methods for measuring the firmness of the ground that are available in modern times, and records show that good well water had already been obtained in a place by the sea—the site of present-day Hama Detached Palace—that at first glance would seem an unlikely place to locate a garden for an estate. These demonstrate that close study of aspects of nature from the perspective of geomancy was always an important factor in urban development. In other words, the relationship of a house to nature and the relationship of a city to nature were considered to be concepts of the same order.

One of the rules of geomancy, particularly in selecting an environment or a site, is that it is infelicitous to live at the bottom of a bluff or at the mouth of a valley. Edo occupied hilly terrain. Important temples and shrines and the estates of *daimyō* (top-ranking samurai) were mostly sited on high ground affording good views, whereas many of the areas for commoners and the residences of low-ranking samurai were located in valleys. The zoning practiced by the authorities clearly reflected the value they placed on their own houses.

Japan, particularly the eastern coast facing the Pacific, is characterized by hot, humid summers. Winds are from the southwest in summer, and the skies are often clear in winter. That is probably one reason samurai estates in early Edo were often long, narrow lots organized around east-west axes and extended in the north-south direction. The Japanese of an earlier time, as pointed out in Chapter 3, not only had respect for climate but sensed the power of place, that is, the presence of the *genius loci* hidden in microtopography. This gave distinctive character to places within the city. In forming a city, they undoubtedly sensed forces not unlike the forces of magnetic fields at work in natural undulations, trees, rivers and landforms. They hoped that such forces could be harnessed in favorable ways, even in developing what was essentially a man-made construct, that is, a community. It is not at all strange that the spirit of geomancy should have many things in common with the principles of urban development. Thus, though a strongly conceptual, systematic character was evident in the early development of Edo, as in the separate modules provided for different social classes, it was always limited in extent to parts of the city. The whole adhered to the demands of geomancy with respect to aspects of nature. The importance attached to the preservation of this balance enabled the Japanese city to achieve a unique urban structure.

In the creation of cities in the West, importance has always been attached to the relationship between the parts and the whole. The parts were conceived to be subordinate to the whole, and if a part was ever emphasized, it was done so in deliberate defiance of the whole. This was true of both the city and architecture. The relationship between the parts and the whole was not perceived in such a way in the process of urban development in Japan. As will be explained in the chapter on *oku*, the Japanese have long seen small spaces as autonomous microcosms and thus developed the perception that a part was in fact also a whole. A single house could be seen as the man-made world in miniature having to contend with all the forces of nature. Everything about Edo from the subdivision of lots and building materials to the overall concept was strictly determined on the one hand, and such microcosms were created at will on the other. A dual psychological structure that was quite unique can be said to have developed in Japan.

However, we should also note that this view of nature, including geomancy, gradually became formalized. That is, nature certainly coexisted with humankind in cities and played an extremely important role in people's lives, but if that were all, a closer relationship between nature and everyday life can be observed in primitive tribes living in the midst of nature. What we are concerned with here is the fact that the various attitudes and ways of looking at things originating in such a view of nature became formalized, in this case through spaces and forms, in a way unique to Japan.

As will be explained in the final chapter, Japanese culture discovered "inwardness" (*okusei*) in mountains covered with broad-leaved evergreen forests. The *torii* gate came into being as a way of endowing the path to inner space with a ritual character, and the *tokonoma* alcove in the *shoin*-style residence—a style of residential architecture that developed in the Muromachi period (1338–1873)—was invented as a symbol of "inner space" itself. The "external layers of a street," discussed in Chapter 4, suggest what nature can be like in boundary domains. In short, when nature is formalized and made into an aesthetic object, it becomes a part of culture.

In his essay "Nature," Munemutsu Yamada,[5] writing on the Japanese view of nature, argues that plants, water and the pail in combination were fundamental to the character of agrarian culture in broad-leaved evergreen forests. Availability of water is naturally one of the prerequisites of a broad-leaved evergreen forest culture, especially a wet-field agrarian culture. Life was lived in the constant presence of water. The sea, the heavens, the verdure of plants, the soaking of yams in water to remove starch, and the pail as a tool used for that purpose—these were combined, and from that combination came rice wine and silk, to which was added tea. Yamada attempts to interpret manifestations of the cultural superstructure such as literature, poetry and art through this basic melding of human consciousness. In short, he maintains that the distinctions found in the aesthetics, philosophies and religions of different peoples can be accounted for by the different ways in which consciousness has been melded.

## STREETS AND BLOCKS

As I wrote earlier, the Japanese attached importance to placeness—the condition of being a place—which is to say they sensed the presence of a force in places where no building had yet been constructed. This is evident in the relationship established between various points of singularity and microtopography (discussed in Chapter 3) but most apparent in the presence of inwardness, discussed in the final chapter. The presence of *oku* helps us begin to understand the unique ways in which the Japanese laid streets and arranged blocks.

If streets are classified from the perspective of what they do—that is, from an operational perspective—within the urban structure, their main functions are to connect, demarcate and arrive. Highways linking two villages or two districts are representative connecting roads. The Japanese word for road, *michi*, is said to have originally been a combination of *mi*, a prefix used to praise the beauty of a thing, and *chi*, a word that vaguely indicated direction.[6] Therefore, in the case of *michi*, directionality was always understood, whether it was, for example, upward or toward nearby mountains. However, this characteristic is not unique to Japan; to connect is the most universal function of roads. Nevertheless, in the case of Japan, roads, even the main highways to Edo, were laid with careful consideration of topography and natural forms. The two ways of laying roads that most obviously conform to natural undulations, that is, along ridges and along valleys, were widely used, and such roads were connected in place by *saka* or sloping roads. A *saka* was not simply a road but a boundary (*sakai*); it represented a domain that included the immediate surrounding area and emphasized the presence of a place (see Chapter 3). Furthermore, there was a hierarchy to ridge roads. In the case of Edo, there were the main ridge roads serving as highways connecting the city to the provinces; then there were ring ridge roads—such as the road from Shiba to the Yotsuya area—connecting the main highways; and finally there were subordinate ridge roads that branched off ridge roads onto plateaus.[7] The fact that this hierarchy was observed in assigning residential areas to various social classes shows the interest planners had in relating the latent power (or value) of land, as evidenced in its aspect, to the expression of authority.

Villages and clusters of residential domains developed along connecting roads. Eventually, back streets developed in addition to the main streets, forming shallow blocks. Any further expansion of the population was absorbed by increasing the depths of the blocks if the configuration of the environment permitted it. This process was typical of villages that developed along highways in Japan.

This form of block development was quite incremental and dynamic. Needless to say, streets in this process of development served two functions, to demarcate blocks and to connect street to street. Streets were laid and blocks formed in a single operation when a grid pattern was used for systematic planning.[8] In Japanese cities, the systematic laying of streets and the simultaneous formation of blocks often took place in the previously mentioned development of towns along highways, or the formation of samurai estates or commoners' districts in castle towns. However, as has already been explained, these were always limited in size and often expressed a clear social hierarchy. That is, cases such as Miletus and Manhattan, where the grid covered an entire city, are rare in Japan aside from ancient capitals. The Japanese assumed a uniformity of power of place only in a relatively level place such as flat, reclaimed land or a plateau and there created a grid pattern of blocks for a group of equal social standing. For that reason, that grid pattern was always limited in scale and its overall orientation freely manipulated in response to various external factors such as topography, relationship to a major highway or, as will be explained later, views. As with streets in other cultures, Japanese streets may have served to demarcate, but in urban development that function was served only within limited areas. Highways may have provided a framework on a larger, regional scale, as can be seen in Edo, the city that eventually became Tokyo, but demarcating streets were used to provide only a part of the whole with a framework. As has already been explained, this was influenced by the extent over which uniformity of space was perceived. As will be discussed in the next chapter on "The Underlying Structure of Streets," a slight displacement where streets converge served to indicate superiority of class of one road over another. When we also take into account the fact that a crossroads functioned as an ersatz public square, we see that intersections emphasized the absence of uniformity over a block or a territory. This will become clear when considered in context.

The relationship between streets and blocks will be discussed later in more detail, but here I would like to point out that in Japan, aside from connecting and dividing, many streets "arrive."

As is explained in detail in Chapter 4, we can still find streets that arrive throughout the city. Among them, the ones that are most ceremonial in character are the approaches to temples and shrines. The prototype is found in the Japanese village situated along a highway, with its approach from the village shrine (*sato no miya*) to the inner shrine (*oku no miya*) in the recesses of a mountain. While its length may differ from case to case, this approach with its ceremonial character has been transposed to cities. Roads that branch off and insinuate themselves into places are most often found in the natural world as in wooded mountains, but in Japan, as discussed in "The Underlying Structure of Streets," they are ubiquitous in what is ostensibly the polar opposite of the natural world, the city. For example, *daimyō* estates on the *yamanote* plateau in Edo were arranged along roads branching off a ridge road onto plateaus. This was a way of signaling arrival. It is similar to the way an approach is laid to a shrine or a temple so that we are made aware of the independent character and inwardness (*okusei*) of the surrounding woods. We know to go no further, that is, we are made conscious of having arrived.

Streets that arrive can be found in even greater number in extremely small spaces in the city. Alleys (*roji*) are typical of such roads. When a site, be it the site of a townhouse or the site of a freestanding house such as a samurai estate, was subdivided, it was not split down the middle—that is, longitudinally—in two. Instead, a narrow cul-de-sac, that is, an alley, was inserted into the site, parallel to a boundary with an adjacent site. In this way, the area behind the preexisting house was accessed. Such a back site was sometimes further subdivided in the same way. Sometimes, houses in an area of *chōnin* were built in *nagaya* (tenement) style, and the alley became a street space that was public in character. This could become even more formalized, as in Kyoto townhouses. Diverse forms of streets developed as house types evolved, including alleys that are like connecting corridors for court houses. These are clearly formalized versions of roads that branch off and arrive. This method of laying streets is of course found even in the West, where a city meets a nearby hill or forest. However, the fact that in Japan such streets are frequently laid near the center of the city, as on the Azabu or Takanawa plateau in Edo, suggests that the Japanese have a unique concept of streets. Certainly, the estate of a lower-ranking samurai or a townhouse had a narrow frontage to begin with, making it difficult to subdivide it into two or three parts. Creating an alley to one side was a simple, easy method of subdivision that did not require the total rebuilding of the house standing in the front portion of the site. However, it was more than just a matter of expediency. From the perspective of territoriality, this way of laying and using streets can be seen to be quite unique in two respects.

First, as has already been explained, Western streets serve first of all to demarcate, and the block formed as a result, no matter what its size, makes possible direct contact between the street and each site in the block.

This is what is regarded as the normal relationship between a street and a block. However, a branch street can be said to impede the establishment of such a relationship. That is because the block is now seen as surrounded by a closed curve, and the principle that any site inside it necessarily faces the surrounding street no longer applies. The way houses in Western cities are numbered is in fact premised on the principle that a street is always in contact with a site. That is, each house is identified by two parameters, the name of the street and a number, just as any point on a plane is identified by values along two coordinates, x and y.

By contrast, the Japanese city, in allowing branch streets to insinuate themselves into territories, does not attach importance to such a relationship between street and block. The second characteristic of branch streets is to be found in the nature of territories and by extension the nature of the outer margins of territories. As can be seen in the relationship between the previously mentioned branch roads leading to *daimyō* estates on the Azabu plateau and the estates themselves, the territory of an estate was delimited by, not a road, but the edge of a bluff. As has already been explained, a Western block is surrounded by a street and can be considered a closed domain. Naturally, a *daimyō* estate on the Azabu plateau abutted other sites at the bottom of the bluff, so that on a larger scale territories including all these sites can be said to have been surrounded in some form on the outside by a number of roads.

However, as will be explained in Chapter 5, a road in that case did not so much demarcate as *wrap themselves around* a territory, and branch roads ought to be seen as similar to nerve ends, leading to territories with extremely ambiguous outer margins. In short, if we look closely at a map showing such an arrangement of streets and the area served by those streets, we discover a territoriality that is unique to Japan, one that was symbolized until recently by the address system.[9] The next task is to examine the exact nature of that territory and the character of the outer margins forming that territory's boundary.

## FIGURE AND GROUND

Let us compare territoriality in the Japanese city with territoriality in cities of other cultures.

If we look again at the Nolli Map of Rome, the city is clearly seen to be composed of two parts, buildings that constitute figure and squares and streets, shown in white, that constitute ground. The first characteristic of territory in Rome is that the exterior walls of buildings, whether those buildings are public or private, constitute the interior walls of public outdoor spaces such as streets and squares. Of course, squares of another kind are at times found on the inside of apartment buildings with continuous facades on the perimeters of blocks, which developed in European cities mainly in the nineteenth century, but in general, the facades of buildings represent a clear boundary between territories for public activities and territories for private activities. However, outdoor spaces such as squares and streets are not the only public places; canopied, interiorized spaces such as arcades and the interiors of churches have long been regarded as public spaces. Thus, the facades of buildings, seen as figure, express the individual identities of buildings but, seen as ground, constitute a part of the continuous interior wall of outdoor space. The elements of the urban typology of early modern times such as gates, walls, squares, streets, arcades and apartment buildings were precisely the elements determining and expressing all territories in cities. Moreover, they were the basic building blocks of the urban proposals advanced in the late twentieth century by Italian Rationalists who sought in them an ethical foundation for urban structure.

The perceived structure of urban territories in the United States is quite different, as pointed out in a comparison of blocks in Paris and Cambridge, Massachusetts, published in *On Streets*, edited by Stanford Anderson of MIT. First of all, in a relatively low-density city such as Cambridge, even apartment buildings are set back from the street, and their walls do not determine the perceived territory of public activity as walls of apartment buildings in Paris (or Rome) do (Fig. 1-4). One of the characteristics of urban structure in the United States is the frequent presence—between streets and buildings—of territories that are private and not intended for public use but belong perceptually to the public domain. This suggests that, semantically speaking, the boundary between public and private is often empty space. American cities

Fig. 1-4 *Left:* Cambridge, Massachusetts, showing overlapping residential territory, public territory and work territory. From Stanford Anderson, *Studies Toward an Ecological Model of the Urban Environment.*
Fig. 1-5 *Right:* Central Manhattan. From Regional Plan Association, *Urban Design Manhattan.*

often give an impression of being hollow, no matter how grand individual buildings may be, particularly when their morphological structure does not correspond to their semantic structure (Fig. 1-5). Perhaps that also accounts for the sense of void in many devastated slum districts, where the structure fails to reflect the intentions or aspirations of residents.

In Europe, such voids are eliminated by pushing buildings to the fore so that they define exterior spaces. However, even where public territories are clearly defined as in Europe, the city as a whole can seem merely like a beautiful ruin if there are no activities in those territories. Such a city does not possess the beauty of a city full of life.

As is explained in Chapter 4, the relationship between the external surfaces of a city and territories within that city is not nearly as clear in Japan as in the West.

First of all, the outer contours of buildings in Japan often do not form the edge of the public domain. The outer contours of buildings are not clear-cut, even where buildings are crowded together. In addition, interstices left between buildings provide access to alleys (Fig. 1-6). Furthermore, hedges or

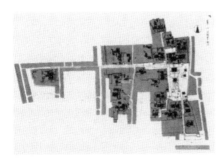

Fig. 1-6 Daitokuji, site plan.

garden walls in many cases clearly divide the public domain from the private in residential districts. Front yards that are perceptually integrated with the street space as in the United States are quite rare in Japan.

In such residential areas in Japan, trees planted inside property walls may grow out over the sidewalk and even the street and become scenic elements in the public space. The houses themselves may only be glimpsed and not fully participate in the streetscape. There are instances therefore where the exterior space is perceptually quite integrated even though territories are clearly divided. This too is a situation that cannot be dealt with by means of Nolli's figure-ground relationship or Anderson's concept of a "domain of public claim."

## GAPS

What are the special characteristics of Japanese urban spaces from the standpoint of morphology and territoriality? First, Japanese urban spaces are characterized by the presence of "gaps." As is explained in Chapter 4, gaps in Japanese cities are not as clearly defined morphologically as ground is in European cities. However, gaps ought to be seen, not as leftover spaces that obscure and possibly vitiate urban spaces, but as mediating spaces that endow urban spaces with a unique tension.

Just as inner space (*oku*) is sensed in places that are hidden from view, as is explained in Chapter 5, a positive meaning is ascribed to gaps within the overall structure.

Gaps in cities have meaning just as intervals between pieces in the game of go have meaning.

In the first stage of Edo's urban development, key elements were arranged around Edo Castle. In due time, areas that developed along highways and *daimyō* estates and the estates of samurai that were built from the start on selected sites gradually expanded.

As is explained in Chapter 2, this form of development was very much like *jintori*, a children's game in which rival groups set up encampments (*jin*) and, using them as bases of operation, proceed to skip about, annexing places. In Edo, gaps were not infrequently left between places annexed by different groups (at least in the initial stage of development). This phenomenon was not limited to the hilly *yamanote* districts of samurai but also occurred in level areas such as the *shitamachi* districts of *chōnin*. Gaps thus produced helped each district establish its own identity, functioning in much the same way as green belts created in English New Towns after World War II. At times they were also what critic Takeo Okuno refers to as "cursed spaces."[10]

These areas, which were more like zones than vacant lots, were to be found here and there around the loop line in Tokyo, at least until about 1935, and seemed like places of mystery and adventure to children hunting for insects. They no doubt derived their power in part from microtopography. Gaps deliberately left were also to be found in the relationship between a site and the house standing on the site, or between a house and the street providing access to the house. Trees, undergrowth, shrubs, hedges, walls and many other elements increased the density and complexity of gaps and contributed to the establishment of new, tense relationships between buildings or between buildings and the street. As the city became more crowded, carefully worked out techniques and devices evolved to deal with gaps of limited size. Of course geomancy was taken into consideration in establishing principles governing relationships between various man-made objects and nature.

Thus the Japanese have regarded gaps not only as places but as boundaries. In short, compared with the hard, one-dimensional boundaries of Western cities, the boundaries of Japanese cities are far softer and more ambiguous.

As Chapter 4 explains, boundaries in Japan are multilayered and multidimensional. I have already pointed out that property walls along streets in an upscale residential district in Japan, unlike those in the United States, allow trees in gardens to grow out so that inside and outside form an integrated landscape, even as they clearly distinguish between what is private and what is public. Territory in this case is defined in more ways than one. Such a boundary is not definite; it is more like an indicator of a sphere of influence. If this were represented graphically, a definite boundary dividing inside from outside would be a thick, unbroken line, and a Japanese boundary would be either a dotted line or a bundle of thin lines. In short, there is not a clear distinction between inside and outside, or between what is on the right and what is on the left. Instead, inside and outside, left and right intermingle. Inside and outside overlap in Japanese houses to begin with, because of the presence of features such as verandas, eaves and *shōji* (sliding screens of translucent paper on wooden lattice). It would not be strange then if the Japanese possessed a similar sensibility in regard to boundaries and perimeters in the broader, urban environment. Gaps and *ma* are products of that sensibility. The fact that Japanese townscapes are stimulating and full of interest despite the often inferior quality of houses and the far from adequate infrastructure is accounted for by not only the delicacy, gentleness and human scale of those townscapes but the sense of tension created in places by gaps and *oku*. Needless to say, plants and natural microtopography play a part in all this. Therefore, any attempt to create for a Japanese city a figure-ground image, such as the one Nolli drew, is made difficult by the fact that there are, in addition to elements expressible as black or white, elements that cannot be expressed graphically; that is, elements that are white but not white. Furthermore, instead of squares, Japanese cities possess *harappa* (open fields). These public spaces, which are not unlike gaps, are symbolic of the special character of Japanese cities.

## CITY WITH MULTIPLE FOCAL POINTS

Japanese cities are often said to have no public squares. They have certainly had few Western-style public spaces other than streets. Moreover, outdoor public spaces in Western cities are often combined with public buildings such as churches, stadiums, markets, theaters and city halls. That means there have also been far fewer public facilities, both in variety and number, in Japan. Edo (the city that became Tokyo) already had a population exceeding one million at the beginning of the nineteenth century, making it the biggest city in the world at the time. Then, London had a population of approximately 860,000, Paris just over 250,000, Vienna 250,000 and Berlin 180,000. Nonetheless, these European cities possessed many public facilities of outstanding quality, whereas the only conspicuous buildings in Edo were Edo Castle, temples and shrines. Even those temples and shrines were situated deep in their respective precincts and rarely participated directly in function or form in urban space. The more important the temple, the deeper it was situated inside its domain in order to underscore the ceremony of arrival. The only buildings that could be called public were theaters and public baths, and from an architectural standpoint they lacked the grandeur to lift the spirit of the public. People engaging in commerce, recreation and urban rituals were all divided by a system of social status, and with few exceptions there was not a demand for such public facilities. Everyday life was lived within a fairly limited area in small social units. Many recreation activities were associated with nature, for example, flower viewing and shellfish gathering. It is often said that Edo was like an enormous collection of villages.

I have hitherto explained the spatial characteristics of Edo, focusing mainly on territoriality as expressed perceptually and physically. It is interesting to consider from the perspective of spatial politics why public facilities and spaces that are so prominent in other cultures were absent in Edo.

Edo was the first city in the world to have a population of one million. The Tokugawa shogunate, based in Edo, was able to maintain peace and order for three hundred years—a rare achievement in world political history.

A feudal organization such as the shogunate considered "divide and rule" a necessary strategy. The *daimyō*, the top-ranking members of the samurai or warrior class that ruled over farmers, artisans and tradesmen under the feudal class system, were obliged to observe the custom of "alternate attendance" (*sankin kōtai*). This enabled the shogunate to control the *daimyō* by requiring them to divide their time between their own domains and Edo (where they left behind their families as permanent hostages). The system worked because Japan occupied an archipelago; it would not have been feasible in a country occupying a continuous body of land. The artisans and merchants who were the residents of Edo were not given public squares or facilities that could become centers of rebellion. Countless places of scenic beauty and historic interest including temples and shrines were arranged instead (Fig. 1-7). There, members of all classes, even samurai of refined taste, enjoyed moments of leisure. In this way, an ingenious strategy of spatial politics—one that

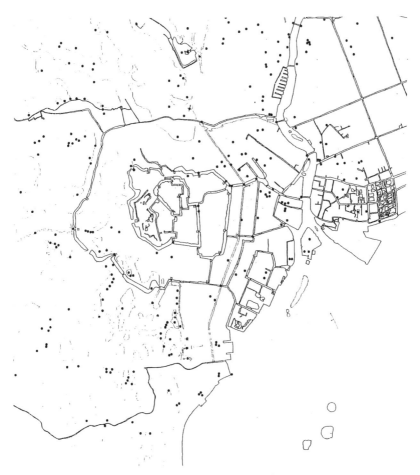

**Fig. 1-7** Distribution of places of scenic beauty and historical interest in Edo.

permitted only small numbers of people to gather—guaranteed peace and order in Edo.

A social structure that is on the whole not highly centralized produces an urban structure that is itself not highly centralized. However, as is explained in Chapters 3 and 5, the absence of a center was offset by the presence of placeness in diverse forms. Many of the names of neighborhoods and places such as slopes, crossroads and fields in Japanese cities are derived from the topography of those places or historical events that occurred there, showing that in their minds and in the actual construction of the urban environment, the Japanese used mnemonic devices and subtle manipulations of the topography to compensate for the absence of a center. Such points of singularity were therefore abundant and diverse. These focal points were by no means monumental but instead integrated into the sphere of everyday life of small groups. Focal points appeared not simply at intersections in town but also on the outskirts.

Examples of this are displacements in street layout, crossroads created through the displacement of intersecting streets, and precincts of small shrines and temples set back slightly from the street. A focal point of this kind gave a district a modicum of identity.

There were a number of stages in the transformation of urban morphology from Edo to Tokyo over the course of four hundred years. At each stage, various principles were made use of in urban development. For each class, residential districts were designated, dimensions of blocks and sizes of lots determined, and building types—*daimyō* estate, samurai estate, lower-ranking samurai estate, townhouse and tenement—assigned. However, as Edo expanded, forms and mechanisms of control intended to reflect the social system collapsed. Since the Meiji Restoration of 1868, Japan has undergone a radical modernization. Edo became Tokyo and new functional, legal and institutional images of the city have been layered over the old city. Moreover, the Great Kanto Earthquake and air strikes during World War II destroyed much of the older urban areas, and what survived has been subject in the decades since the war to both an enormous increase in the volumes and heights of buildings and a fragmentation of property. In addition, suburban districts that were once part of the Musashino plateau and nearby cities are being invested with a new order and purpose. The city has been changing materially as well, from clay, wood and roof tiles to concrete, steel, tiles, plastic and glass.

Notwithstanding those changes, it is not very difficult even today to find traces of the structural principles and styles that have long existed in Japanese cities—methods of urban arrangement based on geomancy; ways of laying streets; ways of treating the external layers of the man-made environment; ways of creating points of singularity; spatial concepts of *oku* and gaps; nature and the formalization of nature that are so closely related to all of the above—not only in the fragments of older urban areas that remain but in newly developed urban areas as well.

To see and understand the city is in fact to understand such structural principles, and we are one step closer to understanding the city as culture when we are able to grasp the city as a thing consisting of two layers: the image that is actually visible and the image that is concealed underneath.

There is one more premise upon which our study has consistently been based. It is our position that decision-making by individuals and groups is constantly at work and affects the outcome in the process of a city's development and transformation. The formation of a city is not like the creation of a single building in that a record of who made a decision for what reason often does not survive. Therefore, we must look to what available records tell us regarding condition A at a certain point in history and condition B after a certain time has elapsed, especially to grasp what principle was followed and for what reason. We must not simply understand what has transpired but discover the underlying principle in the development. A principle must be first presented as a hypothesis, and the validity of that hypothesis must be demonstrated by an analysis of various records. If it cannot be corroborated, it must at the very least be credible. And if it can be discovered that these principles relate to or reinforce one another, then that can demonstrate the existence of a higher cultural context.

That would in fact require a study of tremendous scope. Many of the principles presented in this book have not been completely explained, and such explanations demand more surveys, analyses and thought.

The position we have consistently taken, each time we come across an intriguing historical phenomenon, is to try to clarify why that particular phenomenon occurred in the way that it did. That has been the basic operational proposition, for example, with respect to the laying of streets. We asked ourselves what kinds of streets were created and for what objectives. That is because we believed that the semantic structure generated between the basic methods of operation of streets—namely, connecting, dividing and arriving—and the resulting forms would gradually become clear.

From the standpoint of those doing the creating, architecture or urban development is the performance of a series of operations. For any community in any period, the basic course of action, one that will continue to be followed as long as human society survives, can be described as follows: the establishment of an objective, operation (i.e., concerted activity), result, revision or retainment of the objective, operation, next result and so on.

For us architects and city planners, the difference between dividing a territory and wrapping it is as important a question as the difference between boiling and roasting was for Lévi-Strauss.

Moreover, the more urban development is understood to be, unlike a work of plastic art or a single building, a projection of the cultural will of a group, the greater is the need for such cultural anthropological understanding.

That is because I believe understanding the city as cultural operation or activity provides us with useful clues in developing future cities under entirely different conditions. However, it must be remembered that principles that were effective in the past will by no means be similarly operative under new conditions.

Good examples of this are gaps and inwardness. These spatial concepts unique to Japan retained their meanings in relatively low-density times and places. However, they proved to be a double-edged sword as existing spaces were cut up in cities undergoing rapid modernization. Their emphasis on the structural significance of what is invisible and soft left cities quite vulnerable to external pressures. Gaps and inwardness can also easily vanish with the construction of infrastructure such as expressways and large buildings. On the other hand, in cities that are not highly centralized, they can survive and continue to exert a positive influence, serving as flexible structural principles in small spaces. However, history suggests that when a certain density is reached, a community must discover new principles if flexibility is to be maintained. An historical understanding of the city teaches us that the most important morphological question we are faced with in developing the cities of tomorrow is how to arrive at new autonomous structural and operational principles within the flow of history.

NOTES

1. This suggests that a unique method is needed to express Japanese urban images.
2. Edward T. Hall, *Beyond Culture* (New York: Anchor Books, 1976).
3. Tōkyō no Machi Kenkyūkai, *Toshi wo yomu* (Reading the City), *Toshi jūtaku*, May 1980.
4. Kiyoshi Seike, *Kasō no kagaku* (The Science of Geomancy) (Tokyo: Kōbunsha, 1969).
5. Munemutsu Yamada, "Shizen" ("Nature"), in Masakazu Yamazaki et al, *Nihonjin no biishiki* (The Japanese Aesthetic), (Tokyo: Asahi Shimbunsha, 1974).
6. Munemutsu Yamada, *Michi no bunka* (The Culture of Roads), (Tokyo: Kōdansha, 1979).
7. Tōkyo no Machi Kenkyūkai, *Toshi wo yomu*.
8. See the maps of Miletus and Manhattan in Chapter 5.
9. Japanese streets rarely have names. A district is given a name. Blocks within that district are numbered; thus each block is identified by the district name plus a block number. Individual lots in a block are numbered though not necessarily in sequence; thus each lot is identified by the district name plus a block number plus

a lot number.
10. Takeo Okuno, *Bungaku ni okeru genfūkei* (The Primary Landscape in Japanese Literature) (Tokyo: Shūeisha, 1972).

CHAPTER 2

# THE UNDERLYING STRUCTURE OF STREETS

*Tokihiko Takatani*

## INTRODUCTION

A city is an area overlaid by a network of streets. The circumstances under which streets come into being and the shapes they take are diverse, but if streets are examined collectively, that is, as street patterns, then a number of principles concerning the ways they are made or the relationships that exist between patterns can be grasped from a morphological point of view. In Tokyo there are both areas that were built systematically and areas that came into being virtually spontaneously, and their street patterns differ considerably in appearance. Various studies have already been carried out on the planning principles followed in systematically laid urban areas and the nature of the way streets are laid in areas of sprawl; the focus here has been on finding morphological principles that hold in both types of area. In all likelihood such principles transcend the individual intentions of planners and developers and are deeply related to the nature of spatial awareness shared by people of the same era.

We first undertook an overall examination of the street patterns that were inherited from Edo and became the framework for the street patterns of Tokyo. Next, we selected two or three districts with typical patterns for a more detailed examination and at the same time noted their processes of change. We attempted by these means to abstract a number of morphological principles and to identify their characteristics. The aim was to arrive at an underlying structure inherited by Tokyo from Edo.

## 1. STREET PATTERNS OF EDO

The overall framework of streets in Tokyo was created in the Edo period. The *Kanbun gomaizu*,[1] Edo-period maps known for being based on accurate surveys, correspond surprisingly closely to contemporary maps. The street network of present-day Tokyo has not changed at all in its basic framework from the middle of the seventeenth century.

### THE PROCESS OF FORMATION OF EDO/TOKYO

Edo has a prehistory of development by the Edo family in the twelfth century and service as a castle town in the Muromachi period (1338–1573) under Ōta Dōkan and the Hōjō clan. However, the framework that survives today was formed during the city's development as a political and economic center of the Tokugawa shogunate after Ieyasu entered Edo Castle in 1590. According to Akira Naitō, that process of development can be divided into three stages.[2]

The first stage was the period of construction of Edo,[3] which saw the development of infrastructure for the castle town and the building of a commercial district (quarters for *chōnin* or "townspeople," meaning mainly

Fig. 2-1 The urban area of Edo. From Akira Naitō, *Edo to Edojō*.

artisans and merchants) centered on Nihonbashi and an area for the samurai class surrounding Edo Castle (Fig. 2-1). The urban area of this period is still the center of commercial and business activities today. The second stage lasted from the Great Meireki Fire (1657) to the beginning of Yoshimune's rule (1716) and was the period of completion of Greater Edo.[4] After 60 percent of Edo was destroyed by the Great Meireki Fire, large-scale urban reconstruction was carried out for purposes of fire prevention; at the same time, urban development advanced, for example, to areas beyond the Sumida River. As a result, "Edo was no longer just a castle town in the province of Musashi; a metropolis—Greater Edo (Ōedo)—came into being."[5] The third stage was a period of sprawl.[6] Urban areas were no longer created systematically; instead, as density within existing urban areas increased, sprawl developed on the outskirts of the city. The population of Edo reached "at least 1.3 million,"[7] and the urban area expanded to Sugamo in the north, Naitō Shinjuku in the west, Shinagawa in the south and around Kameido, far beyond Sumida River, in the east.

The system of zoning enforced in urban areas in Edo distinguished three forms of land use: land for the samurai or warrior class, land for temples and shrines, and land for *chōnin*. In the first stage of Edo's development, the zones were each of considerable size and clearly divided from one another, but in the second and third stages, the three zones became more interwoven. Land for the samurai class accounted for 70 percent of the total urban area, whereas land for temples and shrines and land for *chōnin* accounted for 15 percent each.[8] The fact that every *daimyō* had an estate in Edo accounted for the high percentage of urban land occupied by the samurai class.

## CLASSIFICATION OF STREET PATTERNS IN EDO

Having briefly reviewed the history and composition of Edo, let us now examine the city's street patterns.

Edo was made up of not just one type of street pattern but a number of different patterns. Maps make this immediately apparent, but a slightly more objective view is gained by breaking down street patterns into two elements—street shapes and the ways streets intersect—and analyzing those elements. "Street shapes" signifies the configurations of streets between intersections; here, we will distinguish between curved streets and straight streets.

Le Corbusier observed that "The winding road is the Pack-Donkey's Way, the straight road is man's way,"[9] and indeed the latter can be said to be on the whole more rational and systematic than the former. In that sense a street bent into a dog-leg belongs to the category of curved streets.

"The ways streets intersect" signifies the configurations of intersections, and the question is how many streets meet and at what sorts of angle. An intersection of five or more streets was rare in Edo; we will therefore consider only intersections of four streets or less. With respect to the angle of intersection, we will distinguish between perpendicular intersections and all other kinds; that is because a perpendicular intersection, like a straight street, is expressive of a systematic intent of some kind.[10] If there are three possible numbers of streets converging on an intersection—that is, two, three or four—and a distinction is made between perpendicular intersections and non-perpendicular intersections, then six types of intersections in all can be said to exist, but here for practical purposes we will divide intersections into the following three types. First, the cross type is formed by four streets converging perpendicularly. The so-called grid plan is a pattern formed by a set of straight streets with cross-type intersections. Next, the T-type is formed by three converging streets and the L-type by two. Both are perpendicular intersections. L-type intersections ought to be distinguished from dog-legged streets because the former are bent at a right angle.

All non-perpendicular intersections will be lumped together and identified as belonging to a third type. Of the intersections of four, three and two streets, those situations where two streets meet at an angle other than a right angle will be identified as instances of dog-legged streets and included in curved streets. Therefore, non-perpendicular intersections are either four-way intersections or three-way intersections. Thus, street patterns will be described as combinations of two types of street shapes—straight streets and curved streets—and three types of intersections—cross type, T- or L-type and non-perpendicular type (Table 2-1). These types have been plotted on a map of Edo in the middle of the nineteenth century to produce a schematic map of street patterns in Edo (Fig. 2-2).

| Street shape | Diagram | Type |
|---|---|---|
| Straight | \| | Straight street |
| Curved, bent | ( ( | Curved street |

Street shape

| Way streets intersect | | Diagram | Type |
|---|---|---|---|
| Angle | Number of streets | | |
| Perpendicular | 4 | + | Cross |
| Perpendicular | 3, 2 | T L | T, L |
| Non-perpendicular | 4, 3 | + T | Non-perpendicular |

Way streets intersect

Table 2-1 Street shapes and types of intersection.

**Fig. 2-2** Distribution of street patterns in Edo. Based on Yasuo Masai, *Edo no toshiteki tochiriyōzu 1860nen goro*.

## *SHITAMACHI* AND *YAMANOTE* DISTRICTS

As the map makes clear, Edo can be divided by street patterns into two broad regions, the *shitamachi* districts of *chōnin* and the *yamanote* districts of the samurai class. For the purposes of this essay, the largely gridded, low-lying region to the east of an imaginary line drawn north to south from Ueno to Shiba, passing Edo Castle on the east side, will be called the *shitamachi* districts. The region west of this line, where streets were laid further apart and more irregularly, will be called the *yamanote* districts; this region corresponds topographically to the eastern end of the Musashino hills, and the undulating land features many interlocking hills and valleys. The *shitamachi* districts of *chōnin* were centered on urban areas created in the first stage of Edo's development, whereas the *yamanote* districts began to be developed during the second stage.

The *shitamachi* districts were basically gridded districts, but they were covered by not a single grid but a set of small grids of different mesh sizes and orientations (Fig. 2-3). Being two kilometers square, the grid in Honjo on the east bank of the Sumida River was exceptionally large; in areas west of the river, the grids extended about a kilometer in length at the most. The moats and canals that crisscrossed Edo separated grids; a grid was never continued across a waterway. Furthermore, most grids were not perfect grids with only cross-type intersections; there were many T- and L-type intersections as well. T- and L-type intersections as well as non-perpendicular intersections appeared in zones where two different grids met. Grids were arranged around Edo Castle, but a geometrical order was not imposed on that arrangement.

In a feudal castle town, the *chōnin* quarters were usually arranged along an axis perpendicular to a straight line joining the castle to the main gate. In Edo, however, that axis conformed to the lay of the land and was curved like a bow.

In the *yamanote* districts of samurai, there was no dominant pattern comparable to the grid of the *shitamachi* districts (Fig. 2-4). A pattern covered an extremely limited area; even when a pattern of straight streets with cross-type or T- and L-type intersections seems to cover a sizeable area, it is found to contain curved streets.

However, if we examine street configurations in conjunction with topographic maps, then an area with streets that exhibit a certain unity of pattern is found to correspond to a topographical feature such as a plateau or a valley. For example, let us look at the area around Sendagaya on the map showing the distribution of street patterns in Edo (Fig. 2-2). The street patterns are diverse and include a grid plan, straight streets with T- and L-type intersections, and straight streets with non-perpendicular intersections as well

46  City with a Hidden Past

Fig. 2-3 *Shitamachi*.

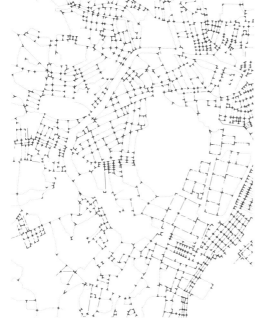

Fig. 2-4 *Yamanote*.

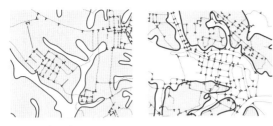

Fig. 2-5 *Left:* Sendagaya area.
Fig. 2-6 *Right:* Kohinata area.

as curved streets. Nevertheless, the streets as a whole are delimited by a narrow, projecting plateau and exhibit unity of character. Similarly, if we examine Figure 2-6, the street patterns, though diverse, suggest an overall unity of character in an area that is topographically a valley.

Thus in the *yamanote* districts of samurai, we can perceive in streets a unity of character that is responsive to topography; the region can be seen as a collection of such unified streets. We shall refer to a set of streets exhibiting coherence of character of this kind a "street pattern unit." Street pattern units were delimited in extent by topographical features such as plateaus and valleys and assumed configurations reflecting those circumstances. For example, the orientation of the streets in the two above-mentioned examples (Figs. 2-5, 2-6) was determined in each case by topographical circumstances.

The small, individual grid patterns in the *shitamachi* districts of *chōnin* can also be called street pattern units in that they can be distinguished from each other. In this way, Edo in both the *yamanote* districts and the *shitamachi* districts was made up of small street pattern units between which no clear geometrical relationship existed. Furthermore, topographical conditions can be said on the whole to have had a great deal to do with their sizes, locations and shapes. The fact that the units were fairly independent in shape and no overall system of control over these units can be discerned creates the impression of an absence of urban order.

## 2. STREET PATTERNS OF DISTRICTS

Next, we will select one district each from the *shitamachi* and the *yamanote* and one district that has only been developed since the Meiji period (1868–

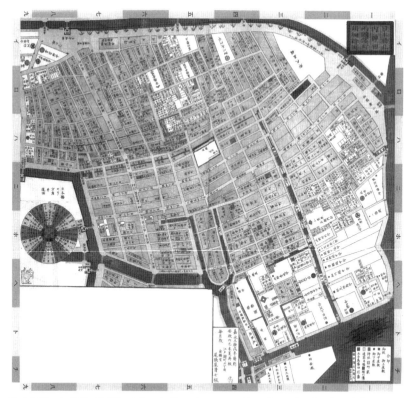

Fig. 2-7 *Nihonbashi Kita Uchi-Kanda Ryōgoku Hamachō meisai ezu*. From *Owariyaban Kaei Keiō Edo kiriezu* in Mikito Ujiie and Takashi Hattori, *Yamakawa MOOK Edo Tokyo kiriezu sanpo*.

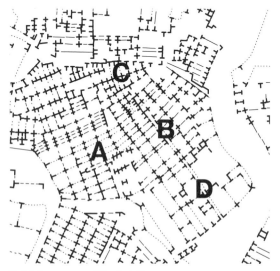

Fig. 2-8 Street pattern in Nihonbashi Kita.

1912) and examine them and their processes of formation more closely, taking into account conditions such as land use.

## NIHONBASHI KITA DISTRICT

The thirty-sheet set of maps in *Edo kiriezu*[11] covers the entire Edo area. Here, we will focus on the Nihonbashi Kita district[12] depicted in one of these maps, the *Nihonbashi Kita Uchi-Kanda Ryōgoku Hamachō meisai ezu* (Fig. 2-7).

This is one of the oldest districts even among the areas first developed in Edo. With respect to street patterns, two grid plans, A and B, occupy the central portion, and patterns C and D characterized by straight streets with T- and L-type intersections appear on the periphery (Fig. 2-8). Whereas A and B are *chōnin* quarters, D is a samurai area that includes the middle residences of *daimyō*, who were exalted members of the samurai class. Samurai residences remain in only a part of C by the end of the Edo period which was when the *kiriezu* map was drawn, but before the Great Meireki Fire C was land for either samurai residences or temples and shrines (Fig. 2-9). After the fire, temples and shrines were moved to the suburbs, to land that had been

**Fig. 2-9** Nihonbashi before the Great Meireki Fire. From *Bushū Toshimagun Edo shōzu.*

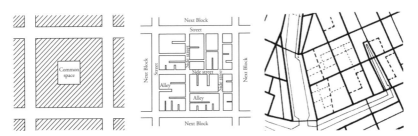

**Fig. 2-10** *Left:* A block in *chōnin* quarters.
**Fig. 2-11** *Center:* A block in *chōnin* quarters, partially revised. From Uzō Nishiyama, *Nihon no sumai.*
**Fig. 2-12** *Right:* Some of the changes in streets.

designated for temples and shrines and samurai residences were subdivided into parcels for *chōnin* houses; however, the street patterns with T- and L-type intersections survived. That is, the difference in street patterns between A and B on the one hand and C and D on the other reflects a difference in original land use.

*Chōnin* quarters created in the earliest period of Edo's history such as districts A and B were in principle made up of blocks such as that shown in Figure 2-10. The shaded area was the part subdivided into *chōnin* houses, and the area left open in the middle of the block "was a common space where latrines and a well for the block were located."[13] However, later, particularly after the Great Meireki Fire, "This portion was given to persons such as merchants serving the shogunate, physicians, artists and Noh actors or leased to merchants for storehouses; eventually ordinary houses came to be constructed, leading to the development of secondary streets[14]" That is, blocks such as the one shown in Figure 2-9 were transformed into blocks such as the one shown in Figure 2-11. A hierarchy of streets developed, with streets (*tōri*), side streets (*yokochō*) and alleys (*roji*).[15] The side streets in a block, as might be expected from the unsystematic way they developed, were rarely aligned with side streets in adjacent blocks (Fig. 2-12). The same was true of alleys.

That is, both side streets and alleys met the main streets in T-type intersections. Though the streets (*tōri*) themselves formed a grid plan, the pattern as a whole was one of straight streets with T- and L-type intersections when side streets and alleys are included.

Such street patterns have remained virtually unchanged through the Meiji and Taishō periods to the present day (Fig. 2-15). Land readjustment was undertaken in this district after the Great Kantō Earthquake in the Taishō period (1912–1926), but with a few exceptions, no changes were made to the

Fig. 2-13 *Left:* An alley (*roji*).
Fig. 2-14 *Right:* A side street (*yokochō*).

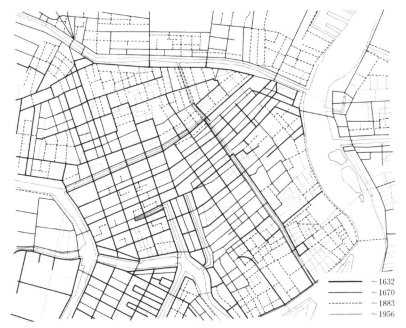

Fig. 2-15 Changes in streets from 1632 Edo to 1956 Tokyo.

configuration of streets. The one conspicuous change was the construction of an arterial road called Shōwa dōri cutting through the district from north to south, but much of this was managed by widening existing streets.

If we shift our focus to the inside of blocks, we find that the organization of streets into streets, side streets and alleys has remained unchanged (Fig. 2-12). However, the side streets, though still of secondary importance, have been widened and changed into places of commercial activity, and the alleys that were once outdoor spaces for tenement residents are now lined with bars, *yakitori* joints and similar establishments that serve office workers unwinding at the end of the day (Figs. 2-13, 2-14).

## AZABU

Next, let us study Azabu, which began to be urbanized in the second stage of Edo's development and is now an upscale residential area. Like the Nihonbashi Kita district, this district is depicted on a map in the *Edo kiriezu* series, in this case the *Tōto Azabu no ezu* (Fig. 2-16).

This district was a small hill surrounded by valleys (Fig. 2-17). A number of roads that cut across the district lead toward the very top of this hill. These

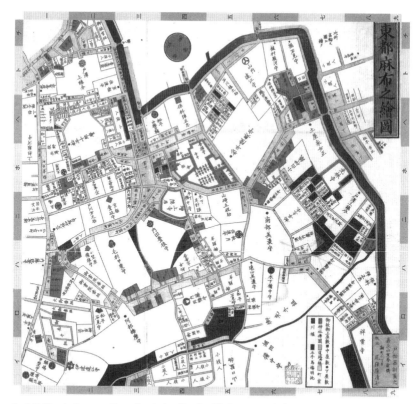

**Fig. 2-16** *Tōto Azabu no ezu*. From *Owariyaban Kaei Keiō Edo kiriezu, 21, Azabu ezu*.

roads follow ridges or valleys and curve gently, reflecting subtle topographical changes. If we examine in the *kiriezu* the way land was used, we find that large samurai estates such as the villas of *daimyō* are arranged along those roads. These estates are all irregularly shaped and bounded by features such as bluffs that create discontinuities of terrain. The rest of the land is divided into smaller, basically rectangular parcels for residences. The roads laid to create these parcels or blocks form patterns of straight streets with T- and L-type intersections. *Chōnin* quarters are to be found along the abovementioned ridge or valley roads. A small community has developed around Zenpukuji; this temple originally faced the road, and the *kiriezu* shows the road-side portion of the temple precinct in the process of becoming a *chōnin* quarter.

How then did the organization of the district and the configuration of streets change into what they are today? From the Meiji period, the population of this residential district continued to increase, and most of the samurai estates were subdivided (Fig. 2-18). However, the ridge and valley

CHAPTER 2  THE UNDERLYING STRUCTURE OF STREETS  53

Fig. 2-17  Street pattern in Azabu. Based on Minatoku Mita Toshokan, *Minatoku henkaku zushū*.

Fig. 2-18  Changes in streets. Based on Minatoku Mita Toshokan, *Minatoku henkaku zushū*.

roads have remained as they were. With respect to the shapes of streets, the ridge roads are curved and have T- and L-type intersections. They have not changed in pattern from the Edo period. By contrast, grid plans and patterns of straight streets with T- and L-type intersections increasingly appear in areas surrounded by the ridge streets. Roads such as the ridge roads that shape the basic framework of the district and the streets that subdivide the samurai residences form contrasting patterns. The organization of the district and the configuration of roads such as those described above, particularly the changes these have undergone since the Meiji period, are typical of what happened to residential areas that were on the outskirts of the city in the Edo period.

## ZŌSHIGAYA

Next, let us examine a district that first began to be urbanized in the Meiji period.

Figure 2-19 shows the map entitled *Zōshigaya Otowa ezu* in the *Edo kiriezu*

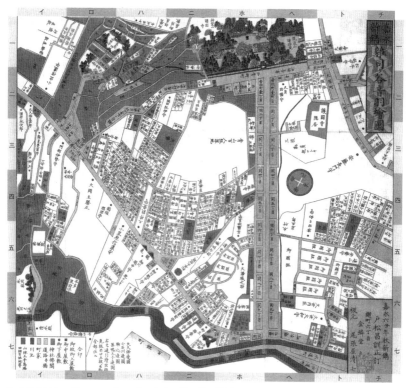

Fig. 2-19 *Zōshigaya Otowa ezu*. From *Owariyaban Kaei Keiō Edo kiriezu, 26, Otowa ezu*.

CHAPTER 2  THE UNDERLYING STRUCTURE OF STREETS   55

Fig. 2-20  Zōshigaya. 1887. From *Naimushō Chirikyoku chikeizu*.

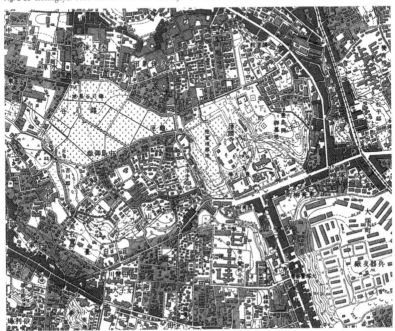

Fig. 2-21  Zōshigaya, 1921. From *Dainihon Teikoku Rikuchi Sokuryōbu ichimanbunnnoichi chikeizu*, Waseda.

Fig. 2-22 Changes in streets.

series. The straight road that runs from top to bottom (north to south) in the middle of the *kiriezu* map is the approach to Gokokuji temple. At the time this was drawn, the Gokoku-ji quarter was at the northwestern end of the urbanized area of Edo, and at the top left corner of the drawing we can see that fields are interspersed among parcels for samurai and *chōnin*. Let us examine the street pattern and the process of formation of that pattern in what is now Zōshigaya, Toshima ward, since the Meiji period.

The area is still farmland on a map of 1887. By 1922, development of parcels has progressed and the streets are virtually what they are today (Fig. 2-21). Many of those streets are based on former field paths (*azemichi*). Figure 2-22 shows the process of formation of streets in a part of the district. The basic framework for the system of streets in the figure was a loop following a contour line, and secondary streets were created inside that loop. In the first stage of development, secondary streets are laid as extensions of the north quadrant of the loop and some of those streets reach the south quadrant. At the same time, streets begin to extend into the interior of the loop, from the south quadrant. Then, streets begin to subdivide areas inside the completed system of streets. The creation of a network of streets based on field paths through such a process is commonplace on the outskirts of Tokyo.

## 3. CHARACTERISTICS OF STREET PATTERNS

We have studied street patterns and their processes of formation in Edo/Tokyo as a whole and in selected districts of the city. A number of characteristics can be abstracted as a result.

Fig. 2-23 Four-street intersections

Fig. 2-24 Intersection in Azabu.

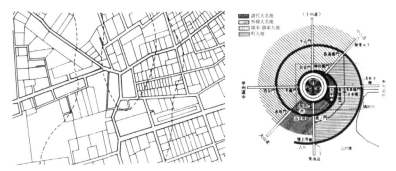

Fig. 2-25 *Left:* Ichigaya.
Fig. 2-27 *Right:* Conceptual drawing of the urban structure of Edo. From Akira Naitō, *Edo to Edojō*.

## ABSENCE OF A CENTRALIZED PATTERN

In Edo a pattern of multiple streets converging on a point did not exist. As was explained in classifying types of intersections and street shapes, an intersection of five or more streets cannot be found. What may appear at first glance to be a five-way intersection is often on closer examination discovered to be a four-way intersection with a fifth street intersecting nearby, and the four-way intersection itself may turn out to be two slightly offset three-way intersections (Fig. 2-23).

Figure 2-24 is a schematic drawing of the main roads of Azabu. As has already been explained, these roads follow ridges or valleys and lead to the top of a hill; as a whole they form a radiating pattern. However, they do not converge on a single point at the top. They instead form three-way intersections. Figure 2-25 shows the main roads in the Ichigaya area in the Edo period. This area is a valley between two hills, and roads gather from four directions toward the valley floor. Here too, roads are offset—and the displacement is so slight as to seem deliberate—so that no more than three streets converge at any intersection.

As mentioned above, a pattern in which five or more streets converge on a

Fig. 2-26 Paris. From Sigfried Giedion, *Space, Time and Architecture*.

single point cannot be found anywhere in Edo, and even when streets gather from four directions because of topographical conditions as in Azabu and Ichigaya, they do not meet at a point.

Research has shown that T-type intersections, staggered crossings and culs-de-sac were originally used for reasons of defense as ways of organizing roads in castle towns including Edo. Not having roads converge on a single point can be understood as another technique of this kind, but such an interpretation, though plausible in the first stage of Edo's history, cannot fully explain the phenomenon in the second and third stages, which were periods of stability for the shogunate. The phenomenon may have to do instead with the nature of Japanese spatial awareness in the creation of forms. If four, let alone five roads were to converge on a single point at the top of a hill or the bottom of a valley as in the above instances, a central place would come into being there. The fact that roads do not converge on a point suggests that the concept of a central place is not something that concerns the Japanese when they organize space (and the act of laying roads is an instance of such organization). Where the concept of a center is weak, roads, even when they come together, are staggered for the slightest reason such as the shape of a preexisting site or topographical conditions. This is in marked contrast to cities such as Paris where Haussmann imposed radiating patterns centered on plazas in a preexisting urban area (Fig. 2-26).

As Naitō has pointed out,[16] Edo as a whole was organized as a spiral in plan (Fig. 2-27). Although Edo Castle was the symbolic center of the city, no particular effort was made to organize the urban space as a whole around the castle; that is, the spiral form can be understood to be the result of the use of existing topographical conditions in creating an urban framework without concern for patterns directly expressive of a center such as concentric or radiating patterns[17].

## GRID PLAN AS AN EXPEDIENT

The *shitamachi* districts were districts distinguished by grid plans. In the *yamanote*, as we have seen in Azabu, streets in areas of small residences for low-level retainers of the shogunate in the Edo period and streets that, from the Meiji period, began to subdivide what had once been *daimyō* estates adopted grid plans.

The grid plan is a highly practical system for producing uniform parcels, but it is also a way of imposing an order based on orthogonal coordinates on the area it covers. A good example of this is the Greek city of Miletus in the fifth century BC.*[18] Although inlets interrupt roads, an order based on the orthogonal coordinate system is imposed over the entire city. The grid plan is the embodiment of a worldview.

A number of grid plans collectively cover the *shitamachi* of Edo, but no such implication can be inferred from them. As was explained in the previous section, the grid plans of Edo were each self-contained and used in an area that was topographically distinct such as a hill or a valley, islanded by canals or intended for a distinct land use. It was rare for a grid to be extended across a canal, however narrow, and continued on the other side (Fig. 2-28).

Furthermore, grid plans broke down and T- and L-type or non-perpendicular intersections appeared at their ends or on their periphery (Fig. 2-28). These grid plans do not seem to have been embodiments of any worldview or ideal. The grid plan was treated as a practical system for producing rectangular parcels in a delimited area, that is, as an expedient.

## *JINTORI*-STYLE APPROACH TO THE DIVISION OF SPACE

The changes that streets have undergone from the Edo period to the present day suggest that the development of urban land proceeded much like *jintori*, a children's game in which players split up into two sides and establish their respective positions or "encampments" (*jin*). Using those positions as bases of operation, they then set out to capture the opponent's position (*jintori*) by annexing places in the irregularly shaped space. In the end, the entire space in which the game is played is filled with a set of occupied positions. Urban land was often formed in a similar fashion. Let us examine some examples of this process.

In the case of Azabu, the *daimyō* estates and temples that occupy land along the ridge and valley roads can be said to have been the first established positions (Fig. 2-29); next, residences of low-level retainers of the shogunate were located in the unoccupied land that remained. The former were irregular

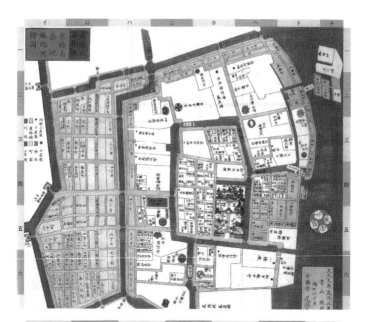

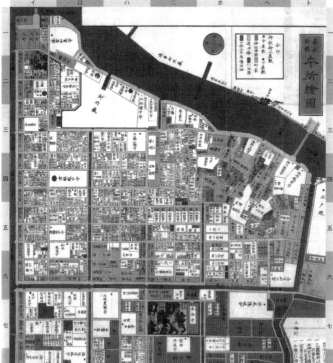

**Fig. 2-28** *Top:* Kyobashi Minami, Tsukiji. From *Owariyaban Kaei Keiō Edo kiriezu*, 7 *Kyobashi Minami Tsukiji ezu*. *Bottom:* Honjo ezu. From *Owariyaban Kaei Keiō Edo kiriezu*, 18 *Honjo ezu*.

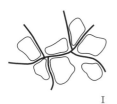  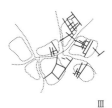

I   II   III

Fig. 2-29 Azabu.

in shape, and the latter were collections of rectangular parcels. With the Meiji period, clusters of small rectangular parcels begin to appear in what had been large estates such as those of *daimyō*. The street pattern was either a grid plan or a plan with straight roads having T- or L-type intersections. Even where, as in Azabu, no dominant pattern can be found and the area seems utterly without order, tracing the stages in this *jintori*-style process of development can help us make sense of the patterns that do exist.

In the case of Zōshigaya, streets were based on field paths that had existed prior to urbanization (Fig. 2-30). The way in which development of parcels proceeded in stages from the north side within the area enclosed by streets was indeed like a *jintori* game. In another district encircled by a street that underwent urbanization after World War II, we can see how land was gradually annexed from the periphery toward the center (Fig. 2-31). In such cases, we find many culs-de-sac.

We have focused up to now on the small scale; if we turn once more to the whole of Edo or Tokyo, we realize that each area with a coherent street pattern, that is, each street pattern unit, represents an established position. Small distinguishable units—street pattern units—are added one after another and reveal no formal interrelationship. This too can be said to be the result of a *jintori*-style approach. The grid plan used as an expedient is an instance of a position established in a *jintori*-style game.

To adopt a *jintori*-style approach to planning is to devise ad hoc solutions,

 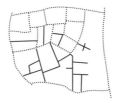 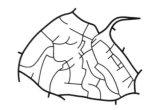

Fig. 2-30 Zoshigaya.   I   II   Fig. 2-31 Kami-Itabashi.

solutions appropriate to situations as they arise within a limited, localized area (for example, a topographical unit such as a hill or a valley). The approach can be criticized as haphazard but does place importance on conditions prevailing on a local level. With this approach, geometry is localized and the whole becomes a random patchwork.

## CITY WITH GAPS

We have focused up to now mainly on areas where streets were laid at close intervals; the very concept of a street pattern unit suggests a high concentration of streets. Let us now turn to areas where streets were more widely scattered.

In the *yamanote* districts, fairly large areas without streets existed between street pattern units. Many of these were large estates such as *daimyō* estates, which made them residential areas and thus parts of the city. These large estates were, as we have seen in Azabu and as a general rule, accorded preferential treatment and located in the best residential environments such as the south slopes of hills. However, the near absence of streets and the irregular shapes of estates create the impression that these areas had no urban land use (Fig. 2-2).

*Daimyō* estates (particularly estates with sunny slopes and access to water) no doubt retained a great deal of greenery, which was used at the very least for gardens. Estates were large, floor areas were small, and natural discontinuities of topography such as bluffs were used as boundaries between properties.

Even when *daimyō* estates were adjacent to, say, a *chōnin* quarter, there was little connection.[19] Thus, we could argue that these areas escaped the urban spatial order even though they existed within the city. These gaps in the network of streets could be said to be non-urban, that is, natural, places.

Similar instances can be found in a *shitamachi* district we previously examined. If we look at the entire Nihonbashi Kita district (Fig. 2-8), zone C can be said to have been a kind of buffer zone generated by the discrepancy between two grid plans, A and B. Zone D can be interpreted as the result of grid plan B breaking down toward the periphery of the plan. Zones C and D were parts of the Nihonbashi Kita district islanded by rivers and canals over which no orderly grid plan extended. Whereas A and B were *chōnin* areas, C and D were areas of samurai residences or temples and shrines, at least when they were originally developed. There, greenery that was absent in *chōnin* areas existed because of the use to which the areas were put. How would this district have appeared from a place that commanded a view such as Kandayama? An orderly grid plan did not cover the entire district, which was surrounded by rivers and canals; instead, greenery remained on the periphery.

The gridded townscape was surrounded by greenery (Fig. 2-32). It was not just the Nihonbashi Kita district; if we examine areas along Sumida River or Tokyo Bay, we see a continuous zone in which street patterns were widely scattered (Fig. 2-2). *Daimyō* villas (*shimoyashiki*) were located in this zone. They were no doubt located there to gain access to water transport, but they effectively protected greenery on the waterside at the edge of the urban area. The *shitamachi* as a whole can be said to have been surrounded by the greenery of the *yamanote* districts to the west and the waterside greenery to the east. In this way, gaps that slipped through the network of streets existed in both the *yamanote* and the *shitamachi* of Edo, and there, greenery remained.

With development beginning mainly in the Meiji period, urban order was gradually imposed on these gaps. However, a *jintori*-style approach to spatial subdivision invariably left gaps between "positions." In particular, places such as bluffs that could not be easily developed retain a natural environment to this day (Fig. 2-33).

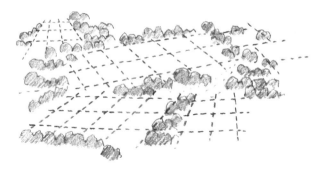

Fig. 2-32  Two views of greenery on the periphery of a district.

Fig. 2-33  Remaining greenery.

## HOMOLOGY WITH RURAL ROADS

Streets in Azabu can be divided into the ridge roads and valley roads that formed the basic framework of the district, and the streets that either created the clusters of small samurai residences or subdivided former samurai estates beginning in the Meiji period. Let us call the former "ridge roads" (*onemichi*) and the latter "streets" (*tōri*). In addition, though we did not deal with them in our previous discussion of street patterns, the alleys (*roji*) are at the bottom of the hierarchy of streets. As we have seen in the Nihonbashi Kita district, and as will be explained in greater detail in Chapter 4, tenements (*nagaya*) were located in the back of townhouses (*machiya*), and alleys provided access to them. Similar forms can be seen in developments that began in the Meiji period.

These three types of streets found in Azabu, though completely different in form (Fig. 2-37), are interrelated in the following way.

First, let us note that the basic function of streets is to connect or to demarcate (Fig. 2-34). A connecting street links two points. By contrast, a demarcating street divides space in two. Actual streets in cities serve both functions, but we can tell which is the main function of a given street.

Ridge roads were not created completely from scratch with the development of Edo but were in all likelihood based on roads, perhaps only rough trails, that had been created naturally in some way. Ridge roads, which carefully correspond to topography, were easily plotted—the lay of the land revealed a way otherwise invisible to the eye for advancing from one point to the next.[20] These can be said to be connecting roads.

Streets (*tōri*) have been a means of creating parcels, for samurai residences in the Edo period and for both public and private sectors from the Meiji period. They demarcate land into individual parcels or blocks. Alleys (*roji*) came into being as gaps, that is, leftover space, between tenements within a

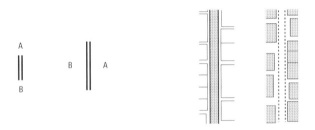

**Fig. 2-34** *Left:* A road that connects and a road that demarcates.
**Fig. 2-35** *Right:* A positive road and a negative road.

block.[21] Even today, they are streets with ambiguous boundaries. It is difficult to discuss alleys from the point of view of connection or demarcation. Alleys are in fact quite different from ridge streets and main streets in being leftover space.[22]

Introducing the idea that streets can be positive or negative space may help explain this point (Fig. 2-35). If, when we imagine a certain district, the street laid there emerges as figure,[23] then that is a positive street. When the relationship of figure and ground is reversed, the street becomes negative. It has been pointed out that in the case of a city such as Rome, a road can be both positive and negative.[24] Such a reversal does not occur in the case of Japan. Alleys which are gaps or leftover spaces in sites can be regarded as negative roads, as opposed to ridge roads and streets which are positive. Figure 2-38 shows schematically this relationship between ridge roads, streets and alleys.

Let us pause here to examine roads in rural communities far from Edo or Tokyo. Figure 2-36 is a map[25] of a village in the Tōhoku region. Its organization seems extremely simple to eyes accustomed to the Tokyo metropolis. The roads that cross in the middle[26] are main roads (*kaidō*) that connect this village to other villages. Along these main roads are farmhouses, and each farmhouse consists of the main building and an auxiliary building.

Between the main building and the auxiliary building is a yard (*niwa*) for farm work. Each main building is accessed from the main road by way of the yard. All around these farmhouses are fields, and there we find field paths. Thus the village is structured by the main roads (*kaidō*), yards (*niwa*) and field

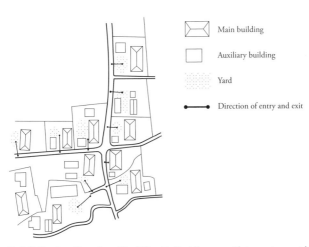

**Fig. 2-36** Farming village. From Hisashi Sasaki, *Kenchiku no gunzōkeiron no tameno ichikōsatsu–kaijō shūraku no bunseki*.

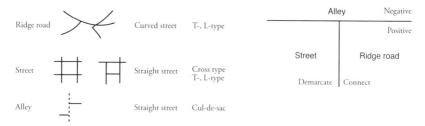

**Fig. 2-37** *Left:* Streets of Azabu (1).
**Fig. 2-38** *Right:* Streets of Azabu (2).

paths.

These three types of roads are different in shape (Fig. 2-39), but let us try analyzing their interrelationship as we did with the roads in Azabu. The main road is a road connecting one village to the next, and the field path is a road that demarcates farmland into small units of cultivation. The yard is leftover space on the site otherwise occupied by the main building and auxiliary building and thus the only negative road among the three. In this way, a schema similar to the one for Azabu can be drawn for the rural village (Fig. 2-40).

If we compare Figures 2-38 and 2-40, we see that the roads of Azabu and roads of the village are homologous in their interrelationship—that is, their structure—within their respective districts, even though their forms do not correspond in every detail. Incidentally, a road structure of the kind found in Azabu is commonplace in the *yamanote* districts. Thus roads of the *yamanote* can be said to be homologous in structure to roads in rural villages. Japanese cities are said to be enormous villages, and this is corroborated by the road structure.

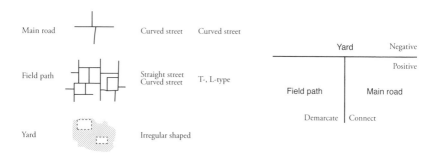

**Fig. 2-39** *Left:* Roads of a village (1).
**Fig. 2-40** *Right:* Roads of a village (2).

## CONCLUSION

The idea of a center is not highly developed; space is developed by a *jintori*-style approach; an appropriate solution is devised for each situation as it arises instead of a single controlling system being imposed over the whole. Gaps between "positions" in *jintori*, or in certain cases entire "positions" slip through the network of roads and become non-urban places—places with natural environments, especially greenery. The grid plan, which in the West was a reflection of cosmic order, was in Edo merely an expedient. The urban area, not subject to systematic manipulation, possesses a road structure copied from that of rural villages. The above are the characteristics of the streets of Edo and Tokyo, and the fact that streets are structural elements of cities is a characteristic of the way Japanese cities are created.

NOTES

1. Collectively referred to as *Kanbun gomaizu*, these maps consist of the *Shinban Edo ōezu* (covering the center of Edo) and the four maps entitled *Shinban Edo sotoezu* (covering areas on the periphery). Compiled from 1670 to 1673.
2. Akira Naitō, *Edo to Edojō* (Edo and Edo Castle) (Tokyo: Kajima Institute Publishing Co., Ltd, 1966).
3. Ibid.
4. Ibid.
5. Ibid.
6. Ibid.
7. Ibid.
8. Ibid.
9. Le Corbusier, *Urbanisme* (Paris: Éditions Crès, 1924), trans. Frederick Etchells, *The City of Tomorrow and Its Planning* (New York: Payson & Clarke, 1929), p. 12.
10. Here too, we are reminded of a statement by Le Corbusier: "The right angle is the essential and sufficient implement of action, because it enables us to determine space with an absolute exactness." Ibid., p. 13.
11. *Owariyaban Oedo kiriezu* (1849–1863).
12. According to Motoaki Tawara, "Kono ningenteki na chizu" (This Human Map) in *Nihon no kochizu* (Tokyo: Kōdansha, 1976), Edo was divided into thirty districts based on "homogeneity of locality reflecting topographical unity and community character," and each district was drawn on a single *kiriezu*.
13. Tetsuo Tamai, "Edo chōninchi no kenkyū" (Study of *Chōnin* Quarters in Edo), *Journal of Architecture and Planning* (Transactions of AIJ), February 1977.
14. Hongō-ku, *Hongō-ku-shi* (History of Hongō Ward), 1937.
15. Various terms including "new street" (*shindō*), "side street" (*yokochō*) and "back street" (*nukeura*) were used in referring to streets, but here all streets surrounding a block are referred to as "streets" (*tōri*), streets subsequently created that pass through a block "side streets" (*yokochō*), and culs-de-sac "alleys" (*roji*).
16. Akira Naitō, *Edo to Edojō*.
17. "This spiral urban form is a truly ingenious conversion of natural topography, using superb civil engineering skills." Ibid.
18. See p.166, Chapter 5.
19. Takehisa Kodera, "Kinsei toshi ni okeru bukeyashiki to machiyashiki," (Samurai Estates and Chōnin Estates in the Early Modern City) in *Nihon kenchiku no tokushitsu, Ōta Hirotarō hakushi kanreki kinen ronbunshū* (Characteristics of Japanese Architecture, Festschrift on the 60th Birthday of Dr. Hirotarō Ōta), (Tokyo: Chūō Kōronsha), 1976.

20. Hermann Schreiber, *The History of Roads: From Amber Route to Motorway* (London: Barrie and Rockliff, 1961).
21. See Figs. 4-27, 4-28.
22. A road at the top of the hierarchy seen from a formal perspective is not always a "connecting" road. For example, consider "Ring Road 6" and "Ring Road 7" in Tokyo. Ring roads are generally planned to save the center of the city from traffic congestion. See Colin Buchanan, *Traffic in Towns* (London: Penguin, 1963). However, as Buchanan notes, they are in most cases not proposals based on surveys of actual traffic trends but intuitive reactions and display little originality. Although this was not Buchanan's intention, this reality signifies the following. Urban ring roads are perceived (albeit unconsciously) by the public not as roads whose most important function is to permit the effective movement of vehicles from one point in the city to another (i.e., to connect), but as linear objects that break down the city into a number of parts (i.e., as roads that demarcate). For that reason, they will probably continue to be built without scientific analyses of traffic volume or clarification of the actual effect of ring roads. As this makes clear, urban ring roads ought to be treated as "demarcating roads."
23. Figure and ground are terms from Gestalt psychology. "Figure refers to a part that stands out clearly and seems unified, and ground the part that seems indistinct and spread out." Megumi Imada, *Gendai no shinrigaku* (Contemporary Psychology) (Tokyo: Iwanami Shoten, 1958).
24. Yoshinobu Ashihara, *Gaibu kūkan no sekkei* (Tokyo: Iwanami Shoten, 1960), published in English as *Exterior Design in Architecture* (New York: Van Nostrand Reinhold Company, 1970).
25. Diverse types of farming villages exist; there is probably no one common type. However, this sort of village is certainly the typical village of the Japanese imagination.
26. The roads are offset slightly in this case as well and forms a three-way intersection. If the roads formed a four-way intersection in the middle of the village, then a center would surely come into being. As is explained in Chapter 3, the symbolic center of a village is located on the outskirts.

CHAPTER 3

# MICROTOPOGRAPHY AND PLACENESS

*Yukitoshi Wakatsuki*

# 1. MICROTOPOGRAPHY AS UNDERDRAWING

## INHERENT CHARACTER OF PLACE

Tokyo is said to be more like an enormous village than a city because it has grown through the assimilation of many villages and is still underdeveloped in its capacity to serve modern urban functions. However, one may take a different perspective: Tokyo could be said to be the way it is because the approach to growth Tokyo and Edo before it have taken is different from modern planning methods, and traces of that process of growth are still visible everywhere. Perhaps the biggest difference of approach is in the understanding of place. In the modern era, nature in the guise of topography or greenery has been mainly regarded as *ground*, a blank slate on which the planner is free to draw *figure*. It is a common belief that all places are equivalent and that the focus of interest in a city is more *figure* than *ground*. As social mobility has increased in recent years, the inherent character of place, that is, *placeness*, has become buried in the continuity of urban space and is becoming ever more difficult to recognize.

Under these circumstances, topography provides an important clue for allowing the inherent character of place to reemerge once more. Japan has many places with a variegated topography. Even in a metropolis, but especially in Tokyo, dissected valleys[1] and uplands interlock in complex ways, and topographical "wrinkles" have obviously influenced the development of communities (Fig. 3-1). Many examples can be found where the power of place is hidden in microtopography,[2] that is, where the presence of a spirit of place as well as special topographical characteristics has been recognized and placeness allowed to ferment. There are even instances where the topography that provided the underdrawing, as it were, prescribed the nature of what was eventually built there. Creating something by employing the latent power of place is an approach that is in marked contrast to the modern planning method, which is to throw off the constraints of place and to situate wherever one wishes. Community development rooted in place cannot be directly translated into contemporary urban development but may still offer many suggestions even today in the sense of compensating for what has hitherto been missing in urban research (and enabling a reexamination of the city from a different vantage point). Adopting such a perspective, this essay will focus mainly on modern Edo/Tokyo and consider its special spatial characteristics through the treatment of microtopography found in community development.

CHAPTER 3  MICROTOPOGRAPHY AND PLACENESS    71

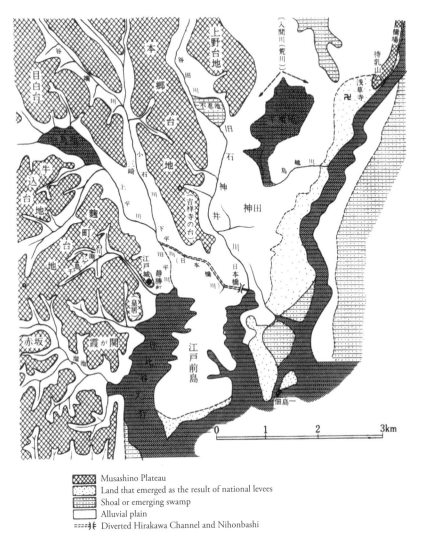

- ▨ Musashino Plateau
- ▦ Land that emerged as the result of national levees
- ▤ Shoal or emerging swamp
- ☐ Alluvial plain
- ╼╾ Diverted Hirakawa Channel and Nihonbashi

Fig. 3-1 The microtopography of Edo. From Suzuki Masao, *Edo no kawa - Tōkyō no kawa*.

## TYPES OF MICROTOPOGRAPHY

A typical Japanese landscape, as most Japanese imagine it, probably consists of houses standing in the middle of fields surrounded by small mountains, with a stream flowing through the fields and a shrine and grove situated on the outskirts of the village. That is the Japanese image of "home"—the mental picture all Japanese share. Such a scene remains embedded in people's consciousness and is superimposed on the contemporary image of a desirable residential environment. The book *Kankyō to bunka* (Environment and Culture) provides a distillation of the ideal environment that today's Japanese visualize: "It is a place of scenic beauty surrounded by mountains, sea and river, on slightly high ground affording a view of the distance. Leafy trees grow around it, seasonal flowers bloom, small birds and animals visit from time to time, and the water and food are excellent."[3] As this description bears out, natural elements in the background of the landscape are an extremely important factor. There is a desirable topography to a desirable environment, and it is seen to be of a certain type. In *The Visual and Spatial Structure of Landscapes*,[4] Tadahiko Higuchi explains the role of topography in the spatial organization of an environment and suggests that the seven following types are representative of Japanese topographical spaces (Fig. 3-2).

1. Sacred mountain type (e.g., Mt. Miwa, Mt. Chausu in Izumo)
2. Domain-viewing mountain type (e.g., Mt. Amanokagu, Mt. Harima-Kunimi)
3. Mikumari Shrine type, a fan-shaped landform where a *kami* or divinity exists that distributes water to villages (e.g., Katsuragi, Yoshino)
4. Secluded valley type (e.g., Hatsuse, Kumano-Nachi)
5. Zōfū-tokusui type (see below) (e.g., Heiankyō, Edo)
6. Akizushima-Yamato type, a basin surrounded by mountains (e.g., Asuka, Mt. Katsuragi)
7. Eight-petal lotus blossom type, a place deep in the mountains, surrounded by a range of mountains that are like the petals of a lotus flower (e.g., Mt. Kōya, Murōji)

These topographies have different meanings but are all delimited by mountains or bodies of water; these landforms suggest a common spatial consciousness.

From ancient times certain topographical characteristics were demanded of sites for capitals. A place bounded by a mountain to the north, hills to the east and west and a body of water to the south was regarded as favorable, and each direction was identified with one of the four guardian gods of a city: Azure Dragon, White Tiger, Vermilion Bird and Black Tortoise.[5] Beyond the mountains and river was another world unprotected by these gods. This ideal was derived from the Chinese system of geomancy known as feng shui[6] used to orient capitals and tombs, and the topography described above was considered a favorable place where the qi[7] of the earth was limited by water and not dispersed by winds. This was a place of *zōfū-tokusui*, where winds were contained and water was acquired, and Edo, which eventually became Tokyo, was also of this type.

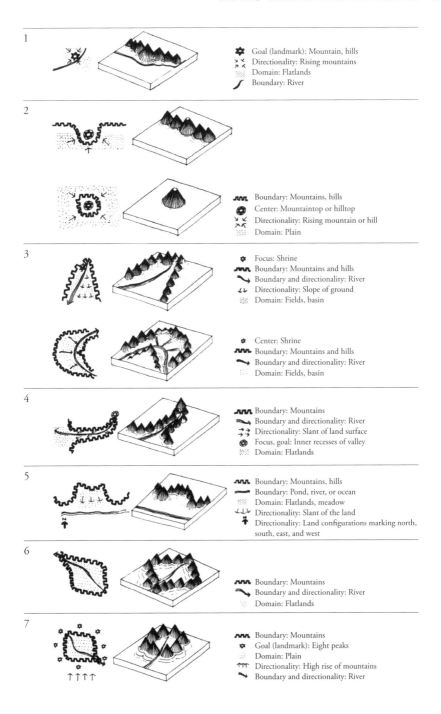

**Fig. 3-2** Types of topographical space. From Tadahiko Higuchi, *Keikan no kōzō*.

# City with a Hidden Past

CHAPTER 3 MICROTOPOGRAPHY AND PLACENESS 75

Divining the fortunes of a city by its physical aspects, like telling people's fortunes by physiognomy or reading palms, seems to have been based on an exquisite sense of the balance—which made it possible to perceive the relatedness—of elements. Such a sensibility was employed in the development of cities and led to the idea of taking into account the preexisting environment, including the topography, and generally respecting preconditions in the creation of things. Such a rule of behavior was subsequently maintained in the planning of temples, shrines, gardens and residences and also became a major factor in the approach taken to community and urban development.

Acknowledgement of the existence of qi in the earth had major implications for the endowment of places with meaning. Even today, though not much consideration is given to the demolition of a building, caution is still displayed with respect to the altering of land. Shinto priests are often asked to perform rites of purification before construction work begins, and a guardian *kami* of one's birthplace (*ubusunagami*) or a so-called "estate *kami*" (*yashikigami*)[8] who has been displaced by the construction of an office building is often enshrined on the roof of the building. An unruly *kami* is thus placated and asked to become the guardian *kami* of the place in question. Such *kami* were believed to wield power over certain places, but the scale of their influence varied. The presence of diverse *kami* was acknowledged—their spheres of influence could range from a single lot to an entire city—and formed the basic order among heterogeneous places within urban space.

## 2. COHERENCE AND EDGE IN MICROTOPOGRAPHY

### THE TERRITORIES OF *KIRIEZU*

Looking at maps known as *kiriezu* from the Edo period, one discovers that the districts depicted are variously sized and the scale used is different from one map to the next. "The maps were manipulated so that a district with a coherent topography and distinctive character would fit on one, easily folded, rectangular sheet of paper."[9] Until the Meiji period, people's everyday lives took place within one district, which encompassed everything from dwellings and workplaces to places of play and entertainment.[10] It is worth noting that people living in relatively small districts in close proximity to one another within Tokyo were believed to have different temperaments, as witness such terms as Kandakko, Shibakko and Fukagawakko signifying respectively

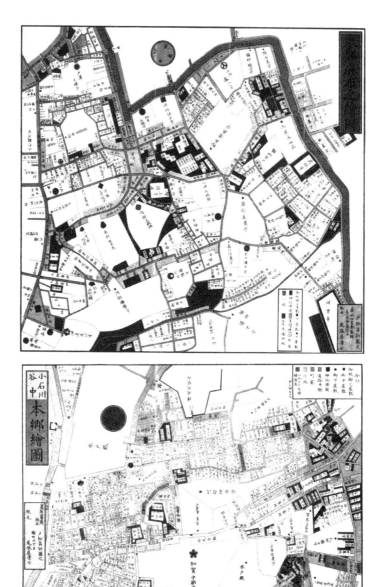

**Fig. 3-3** *Kiriezu*, district maps, Azabu (above), Hongō (below). From *Owariyaban Kaei Keio Edo kiriezu*.

persons born and bred in the Kanda, Shiba and Fukagawa districts.

Topography is not shown on *kiriezu*. However, if *kiriezu* maps are compared with the *Sanbō honbu sokuryōzu* (Army General Staff Office Survey Maps) of 1888 and *Tōkyō jibanzu* (Tokyo Ground Maps) of 1902, one finds, for example, that the territory that is called Azabu in *Tōto Azabu no ezu* (Drawing of Azabu in the Eastern Capital)(Fig. 3-3) corresponded to a peninsula-like upland bounded to the south and east by the river Furukawa, to the west by a small stream flowing through a valley stretching from Tengenji to Kōgaichō, and to the north by a road running along the ridge of a hill from Roppongi to Iikura. Another territory called Hongō is discovered to be an area organized around the Nakasendō, a highway that runs along ridges, and bounded by three rivers, Kanda, Aisome, which flows through Nezu valley and into Shinobazu Pond, and old Koishi River, which flows through Sasuga valley. If one looks at the city as a whole, Edo was made up of territories drawn on twenty-eight *kiriezu* maps corresponding to microtopographic folds created by hills approximately 60 *shaku* (18m) in height and a water system that included rivers and canals.[11]

This correspondence between districts and topography can be found even on a small scale. According to a *kiriezu* of the Azabu Jūban district (Fig. 3-4), many of the areas for *chōnin* (townspeople, mostly artisans and merchants) on the upland were found along ridge roads. When samurai residences dispersed from the center of Edo to surrounding uplands, the *chōnin* who provided Katamachi (services to the samurai class) followed, as in Iikura-Katamachi. Whereas *chōnin* areas on ridges were linearly organized and limited to a single row of buildings on either side of a road, those in valleys were spread out and delimited by topography, as for example in the area from Higakubo to Azabu Jūban. The place names of the latter type often reflect their low-lying

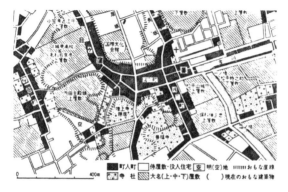

Fig. 3-4 Roppongi and Azabu Jūban in late Edo period, according to an Edo *kiriezu* of 1861.

locations; for example, *kubo* in Higakubo means "hollow," and Miyashita and Sakashita are literally "below the shrine" and "bottom of the hill." The territories in valleys were adapted to microtopographies and moreover divided into small blocks; they were therefore very different from *chōnin* areas on higher ground in street pattern as well. Samurai residences and major temples and shrines, on the other hand, were often built to make use of the "wrinkles" of microtopography on uplands. There were of course considerable differences among members of the same samurai class. The high-ranking *daimyō* had large estates on the edges of uplands, and the smaller residences of low-level retainers were located on lower ground; thus, status was directly expressed by position in the topography as well as size of residence.

A *daimyō* residence in its extent closely corresponded to a topographically coherent area (Fig. 3-5). According to the Hongō *kiriezu*, the middle estate of Abe Iyo no Kami was near what is now Nishikata in Hongō; when this is compared to a topographic map, the estate is found to neatly match a spur of high ground. Such places remained areas of coherent character even when they later underwent residential development. There are many instances such as Nishikata of a samurai estate being transformed into a quiet, exclusively residential district (Fig. 3-6). Meanwhile, the low-lying area with its relatively smaller sites became a more densely built up district with mixed commercial and industrial uses as well as residential uses. The city of Edo thus had a dual system of zoning. There was a large, urban system of zoning that distinguished between diluvial uplands (*yamanote*) and lands reclaimed in the Edo period (*shitamachi*), and a smaller, more localized system of zoning corresponding to microtopographic wrinkles.

## NODES OF ROADS

Microtopography was always taken into account in the way roads were laid in Edo. In *yamanote* areas, as Chapter 2 makes clear, roads following the lines of ridges and valleys generally formed a framework from which secondary roads branched off to cover uplands and valleys (Figs. 3-7, 3-8). For this reason, roads on the whole were discontinuous, and the names of roads were also localized. Even today, one will be traveling on a road and discover after a while that it now goes by an altogether different name. The configuration of roads in Tokyo does not lend itself to an understanding of the city as a whole. A place in a city with streets laid on a grid such as Kyoto can be identified by two intersecting streets, but in Edo, in all likelihood because roads followed microtopography, a place was identified by a point. As can be seen in the case of Surugadai (Fig. 3-8), the street pattern on an upland becomes irregular on

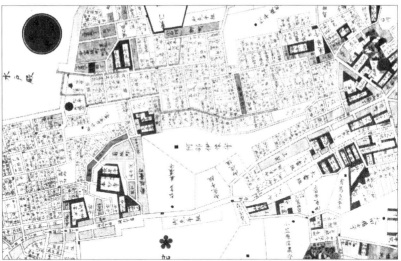

**Fig. 3-5** *Top:* Topographic map of Hongō, Nishikata. *Bottom:* Edo map showing the corresponding area, from *Owariyaban Kaei Keio Edo kiriezu, 14 Hongo Yushima ezu.*

the periphery and diverges from the pattern in an area below it. Such points of divergence are apt to disorient, but emphasizing those very points can in fact make a place more striking and memorable. A hill road where the street pattern changes is an important feature from the point of view of spatial awareness; the street forms the edge of a territory.

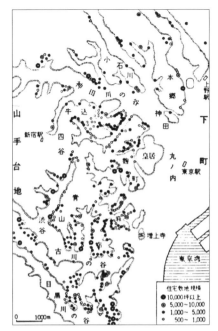

Fig. 3-6 Size of residential sites in areas above *saka* in eastern *yamanote* district. From Nihon Chishi Kenkyujo, ed., *Nihon Chishi 7*.

Fig. 3-7 *Left:* Branching roads in Azabu.
Fig. 3-8 *Right:* Grids on uplands: Surugadai.

## HILLS (*SAKA*) AND BOUNDARIES (*SAKAI*)

Roads in Edo may have a district name but no numbers. Since the enormous samurai estates, temples and shrines that were scattered throughout the city did not even have district names,[12] hills were convenient landmarks. With regard to hills (*saka*) in Edo, Eichi Yokoseki writes, "Hills were the only landmarks in Edo. In particular, hills were used to explain the location or scale of major fires in the city. For example, a fire might be described as having started in Gyōninzaka in Meguro and to have destroyed everything from Kikuzaka in Hongō to the neighborhood of Dangozaka in Sendagi. After a major fire in Edo, hills were the only things remaining that were visible even from a distance."[13]

There are many hills in Tokyo, and almost all of them have names (Fig. 3-9). Only about four hundred well-known *saka* exist, but there are seven hundred *saka* names; that is, names outnumber readily identifiable hills. *Saka* names, moreover, have long histories, and 70 percent of all *saka* have existed since the Edo period, as attested to by, for example, inscriptions on *kiriezu* maps. This phenomenon can be said to be distinctly Japanese. San Francisco is known for its many hills, but city grids are laid geometrically with little relationship to its micro-topography. A hill is always a part of a street; the

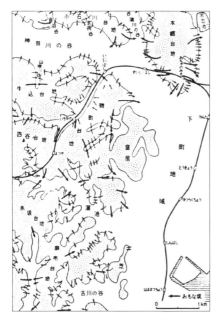

**Fig. 3-9** *Saka* map in Tokyo. From *Nihon Chishi 7*.

street may have a name, but the hill portion of the road is not given a name of its own.

Names that include *saka* indicate one of six things.[14]
1. A topographic condition (e.g., Dangozaka)
2. The natural landscape (e.g., Fujimizaka, Hinashizaka)
3. The shape of the hill (e.g., Yoroizaka)
4. A legend or tradition (e.g., Nekomatazaka, Tanukizaka)
5. The name of a temple or shrine (e.g., Fudōzaka, Tsumakoizaka)
6. A samurai family name (e.g., Hattorizaka, Ikizaka)

These names all evoke a specific image or narrative and reinforce the significance of the place as a transition point. *Saka* names that relate to topography remain effective and easily understood guideposts even today. The many Fujimizaka and Shiomizaka found throughout Tokyo may no longer afford views of Mt. Fuji or Edo harbor as their names suggest, but they continue to indicate the orientation of the *saka*. (Fujimizaka and Shiomizaka are respectively "hill with a view of Fuji" and "hill with a view of the sea") Names given to steep hills such as Munatsukizaka ("chest-touches-ground hill"), Haizaka ("crawl hill") and Korobizaka ("tumble-down hill") emphasize the actual physical experience of climbing up or down the hill. A name may also indicate the overall topography of an area. A long, narrow valley leads westward from Akasaka-mitsuke, and a road cutting across it goes downhill and then uphill (Fig. 3-10). One name, Yagenzaka ("druggist's mortar hill"), is used for the pair of slopes (Fig. 3-11). When a road crosses a hill, on the other hand, the slope is said to split (*waru*) an earthen pot (*nabe*) turned upside down; hence, the hill heading toward Hirakawachō past the National Theater in Hayabusachō is called Nabewarizaka. By contrast, Utou-zaka in Ōji-honmachi, like Utaizaka and Utazaka, indicates that the *saka* cuts across the tip of a rise on otherwise level land. Nezumizaka ("rat hill"), which leads from Azabudai toward Mamianachō has a name suggesting the opposite situation;

**Fig. 3-10** Present-day Yagenzaka.

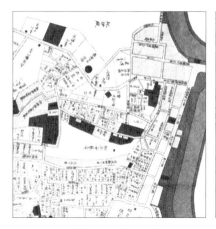 

Fig. 3-11 Yagenzaka: *kiriezu* and topographic map.

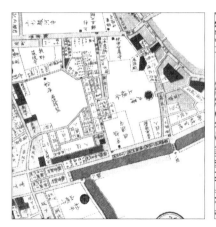 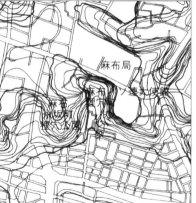

Fig. 3-12 Nezumizaka: *kiriezu* and topographic map; a landform suggestive of a valley.

the road descends into a dead end closed in on both sides by hills (Fig. 3-12). These various names indicate the configuration of the topography and the sense of coherence of place produced by that configuration; they indicate one's relationship or orientation with respect to topography and suggest that the point at which the hill begins to rise or fall is the entrance to or the boundary of the territory lying ahead. Such *saka* names respond to certain demarcations of territory and evocatively express the character of the places thus defined.

There seems to be a connection between *saka* and the word *sakai*, that is, boundary. A *saka* is the point of transition from low-lying land to upland and traditionally has been regarded as a territorial edge. There is a poem in the *Man'yōshū*, Japan's ancient collection of poetry,[15] in which the expression "*unasaka o koete kogiyukuni*" ("rowing past the *saka* of the sea") is used. *Unasaka* meant a boundary at the far end of the sea. In Japan's ancient myths, Yomotsuhirasaka, the place where Izanagi (the male deity) obstructs the pursuing Izanami (the female deity) with a boulder, is considered the boundary between this world and the world of the dead. A hill was thus a boundary between two different worlds and by custom considered important as a strategic point from which the world on this side might be ruled. Pillars of wood or stone or large trees were often placed and worshiped at hills— for example, those dedicated to Sumisaka no kami and Ohosaka no kami. Placing such an object merely made tangible the meaning of a hill. The *kami* that ruled the *saka* was not worshiped; instead, the *saka* itself was worshiped as a *kami*.[16] The act of worship safeguarded that strategic point, politically and psychologically, and helped maintain order in the world within. Traces of such beliefs can still be found in local communities. As can be seen in *saka-mukae*, the custom of meeting a person returning home from a journey at the hill leading to the village, hills and mountain passes formed another boundary beyond the outermost houses of the village. For that reason, a hill was regarded by custom as a territorial edge; there, marriages between families from different villages were discussed, and barter markets were held since each side needed to transport goods only half the distance.[17] Being the threshold to a community, a hill was a place where stone pillars and other markers known as *sae no kami* were worshiped and roads were often barred by *shimenawa* (sacred straw festoons) to prevent the intrusion of plague, smallpox and evil spirits and demons in general.

## *SAE NO KAMI*

*Sae no kami* are found in various places even in cities. Inari (the Japanese deity of fertility and rice), Jizō (the bodhisattva Kṣitigarbha), *dōsojin* (the guardian deity of travelers) and *kōshinzuka* (a pillar or monument dedicated to the Tantric deity Kōshin Shōmen, also known as Shōmen Kongō) were placed near bridges and hills, crossroads, alley entrances and corners of estates and served to obstruct and drive away misfortune (Fig. 3-13). They were often located at forks in roads and other intersections. A fork in the road is a three-way junction and the most common form of intersection in Tokyo. A fork is often found in places where the road pattern is disturbed by a transition in grade or where a road bifurcates to circle around a hill. Deities are said to have originally been worshiped at a fork in the road to assure rich crops, but as the name *dōsojin* (literally, "road ancestor *kami*") suggests, forks in roads also seem to have been regarded as entrances to unknown worlds. Forks, which are

Fig. 3-13 Inari shrine at the entrance to an alley.
Fig. 3-14 Dōsojin at a fork in the road.

also landmarks, were not only provided with guideposts but had little shrines or halls for stone deities or buddhas that were worshiped as *chimata no kami* (*kami* of forks) who protected travelers (Fig. 3-14). Shrines located at a fork often face the corner of the intersection, and such an arrangement is quite rare for Japanese houses. The gate to a house at the end of the road at a T intersection is ordinarily displaced from that road, and in its stead a statue of Shōki (a Taoist immortal) or a *shiisa* (an Okinawan lion dog) is placed to ward off misfortune. When a Shinto shrine is situated at a fork in the road, it almost invariably faces the tip of the three-way junction (Fig. 3-15). In the case of Kumano Shrine, the *torii* gate faces the three-way junction and the approach is laid to split the fork in the road. The road leading to the *torii* is an extension of the approach and indicates the route to the shrine; the roads branching

Fig. 3-15 *Left:* Hikawa Shrine facing a three-way fork.
Fig. 3-15 *Right:* Kumano Shrine and its relationship to a three-way fork.

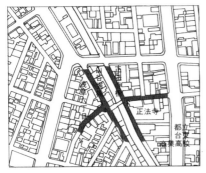

Fig. 3-16 Shōhōji Bridge in Asakusa.

Fig. 3-17 Shōhōji Bridge and shrines at its ends (*left and right*).

off to the left and the right signify a departure from the main approach. Depending on which way the shrine faces, the three ways are no longer equal; the territory becomes divided into this side and the other side, where the road forks. In the case of Shōhōji Bridge in Asakusa (Figs. 3-16, 3-17), this form of obstruction is found within the city. The canal known as Sanya-bori over which Shōhōji Bridge crosses forms a boundary, and the two sides are different both in their history of development and in their planning. Although the two sides of the canal are connected by a bridge and both have a shrine where the bridge meets the road, they are semantically divided by *sae no kami*.

Certain patterns are followed at boundaries of towns, depending on the relationship between the object placed and the place in question. Japanese hackberry is often planted in a mound at a fork in the road; and at temples on the outskirts of town, a pine that looks like a person bent over crying (*yonaki-matsu*) was considered more appropriate than a straight-standing cedar. Formulas were established for places in towns; for example, a Koyasu Jizō, said to assure safe childbirth, was situated on a riverbank, and an Inari at the entrance to an alley. Certain *sae no kami* came to be associated with points of singularity such as hills, cliffs, riverbanks, forks in roads, alley entrances and outskirts of town; formulas came to be established and maintained for Japanese thresholds. Most *sae no kami* were rooted in place and not all that powerful in delimiting space. They did not enclose territories; instead, a number of thresholds were arranged like *go* pieces and created *kekkai*-like boundaries[18]; the result was a finely-textured space responsive to diverse scales.

## 3. MICROTOPOGRAPHY AND TOWN PLANNING

### LANDSCAPES WITH VIEWS OF HEIGHTS

Mountains pleasing to the eye that were visible from town were regarded as the territory's outer edge. Mt. Fuji, Mt. Tsukuba, Sumida River and Musashino were regarded as the four best-known scenic features of Kan'hasshū[19] from the early Edo period. The vast Kantō region was already perceived by then as one space surrounded on three sides by mountains and facing the sea to the east and the south. The surrounding mountains were important landmarks for understanding relationships between places. As can be seen in the *Edo meisho zue*, a guide to places of scenic beauty and historic interest in Edo, Mt. Fuji and Mt. Tsukuba were depicted in the distance in many of the drawings in the Edo-period guidebook to the city. In the drawing of Surugachō (Fig. 3-18), Mt. Fuji was drawn at the far end of the streetscape; it played a major, symbolic role, reinforcing the identity of the district organized around this street.

"Borrowed scenery" is a landscaping technique long used in Japan in which a view of a distant mountain is incorporated into the garden design. The view of the mountain is not simply a background; it becomes an element in the composition and creates a scene with a dynamic perspective. This garden-design sensibility may have informed the planning of urban areas in Edo. According to Shinjirō Kirishiki, "Instead of making use of natural topography, the *shitamachi* areas of Edo were built on land created through large-scale reclamation. The blocks could have been laid in an orderly fashion, but that was not the case. In particular, there seems to have been no rule to the way the

Fig. 3-18 Surugachō and Mt. Fuji, as depicted in woodblock prints.

main street of an area was arranged; no formula seems to have existed for the orientation of the main street or the relationship of that street to other major streets. However, there is a clear solution to this riddle. A three-dimensional point of view was adopted, and the city was planned so as to provide views of heights from low-lying areas."[20] In his opinion, the axes of major streets were set so that Mt. Fuji and the heights in Edo were visible.

If one looks at the organization of the central area of Edo (Fig. 3-19), Tōrichō, the street from Kyōbashi to Nihonbashi, is oriented toward Mt. Tsukuba; at Nihonbashi, it changes direction, and the street until Sujikai Bridge has a view of Yushimadai. As the drawing in the *Edo meisho zue* has shown, Honchō dōri, a street intersecting Tōrichō, is oriented toward Mt. Fuji. Daimyō kōji, the street just west of Tōrichō, is oriented toward Shinobugaoka in Ueno. Numerous other examples of views of mountains or other heights terminating street axes exist, not only on reclaimed land, but in delimited grid-planned upland areas that have been previously mentioned.[21] Other such examples include the streets in the Banchō district that look toward Mt. Fuji or Kandayama; Ichigaya Nandomachi (the site of Fujimi Riding Ground) which looks toward Fuji; Shiba Shinmeichō, which looks toward Zōjōji Maruyama; and the street from Kayamachi to Kuramaekatamachi, which

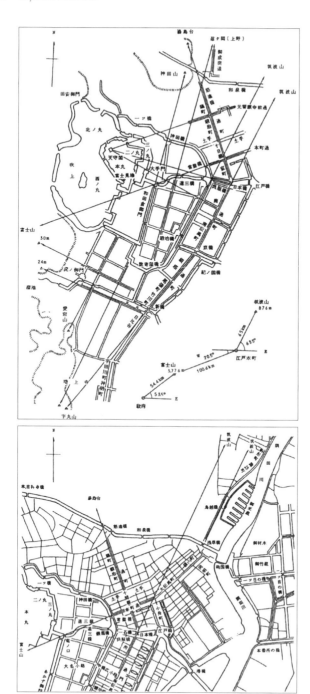

**Fig. 3-19** *Top:* Urbanized area in early Edo. *Bottom:* Organization of the central portion of Edo in the early to mid-17th century. From Shinjiro Kirishiki.

looks toward Mt. Tsukuba. Areas named Fujimichō (literally, "Fuji-viewing town") still exist in many places. As these examples show, it was not just tall mountains that served as focal points; bluffs and knolls such as Maruyama, Atagoyama, Kandayama, Yushimadai and Shinobugaoka were also chosen. Because they were closer to *shitamachi* areas, their angles of elevation were comparable to those for the much taller Fuji and Tsukuba.[22] Many more street axes are concentrated on Kandayama, Yushimadai and Maruyama where major temples and shrines stand than on other knolls, suggesting that they were considered specially important not only as strategic points of defense but also as planning data.

### DISPLACEMENT OF PLAN

Places where the street axis shifted and the focal point in the townscape changed were nodes in Edo. Checks to see where such transition points were in the previous organizational drawings show that many of them were bridges such as Nihonbashi, Kyōbashi, Shinbashi and Gofukubashi or *mitsukemon* such as Sujikaigomon, Tayasugomon and Ichigayagomon. *Mitsukemon* were gates equipped with guardhouses, and in times of emergency bridges could be destroyed to better defend areas inside castle gates. However, as time passed, the defensive purpose of both bridges and *mitsukemon* became less important; for example, at Sujikaigomon, an open space called Yatsutsujigahara was created and became a lively site for shops. The end of a bridge also became an area of diverse activity—the Japanese equivalent of a square or plaza— with a vegetable market, a notice board where the shogunate posted official announcements, a landing place, a guardhouse and *tokomise*.[23]

    The portion of a city defined by a bridge was often at odds in its planning with adjacent areas. It was not just a difference in orientation. The bridge provided a connection, but there was apt to be no continuity of way. Thus, the bridge resembled a *saka* in the *yamanote* district in that the displacement became a boundary; it was probably perceived as such. That is, a bridge (*hashi*) was also an end or edge (likewise *hashi* though written with a different *kanji*). Even reclaimed areas of Edo laid out in a grid were delimited; canals divided them into territories, each with its own plan. For this reason, the grid in Edo was fundamentally different from an infinitely extendable ideal grid. Edo was made up of many fragmentary grids, each complete in itself, with different orientations. One of the reasons for this was probably the fact that mountains and lesser heights were made the data for planning. In the case of Rome or Florence (Fig. 3-20), the focal points of axes lay within the city and, moreover, were man-made structures such as churches, towers or obelisks; the intention

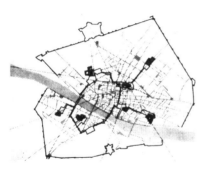

**Fig. 3-20** Florence. From Edmund N. Bacon, *Design of Cities*.

was to organically relate those points. The network of axes was extended over the entire city through conceptual manipulation; the parts of the city were organized with a more rigorous geometrical order. In addition, in many cases the focal points exerted a more powerful centralizing force. In the case of Edo, the castle was clearly a focal point for the city, and one can recognize in the city plan an overall spiral structure with the castle at the center, created by means of the five main highways and canals. However, highways were easily bent by microtopography, axes of streets were displaced, and focal points did not exert a powerful centralizing force. That is, Edo Castle exerted a hypothetical force over the overall, general structure of the city, but that organization was not carried through in the parts. Each district was planned with its own focal point; those focal points, moreover, were already existing natural features such as mountains and knolls on the outer edges of districts. From looking at its plan, one cannot help but get the impression at first glance that the city was the result of "natural" growth. Nevertheless, as has already been shown, if a three-dimensional point of view is adopted, one sees that existing natural features were incorporated into the townscape as focal points and that, as with "borrowed scenery," linking such focal points to places strengthened the character or identity of those places. Whereas in Florence the city was made to embody an idea, the tendency in Edo was to create the townscape according to rules or conventions based on a spatial sensibility honed in garden design and castle planning.[24] Instead of applying one principle throughout the city, priority was placed on acknowledging circumstances prevailing in a particular place. For that reason, the city of Edo seems the expression, not of the strong will of an urban planner working at the behest of the shogunate, but of group intentions—to put it another way, a product of collective norms expressed over an extended period of time. As focal points have been eliminated by the construction of tall buildings and the townscape has changed, the city has come to seem without order. The relationship that characterized Edo between an overall system that was open and loosely defined and parts that were autonomous was tenable only when careful consideration was given to each place.

## HEIGHTS, TEMPLES AND SHRINES

The major temples and shrines of Edo were located above bluffs that served as focal points of townscapes.

Temples and shrines patronized by the shogunate such as Zōjōji, Atago Shrine, Sannō Shrine, Kanda Shrine and Kan'eiji were constructed above bluffs so that "the red shrines and the tiles of temples, which stood out against the background of greenery, could be seen from the areas below,"[25] thus reinforcing their significance as focal points. By the same token, nearly all the precincts of these temples and shrines became well known for their views of Edo and attracted crowds in the manner of parks much as the ends of bridges did (Fig. 3-21). It has been pointed out that the old Inari shrines of Edo were all located above bluffs.[26] Bluffs were on the periphery of early Edo, and temples and shrines such as those dedicated to Inari are believed to have defined the boundaries of the city. Climbing a hill and praying for safety at a shrine must have made palpable the act of leaving Edo.

There were roughly speaking four types of location for temples and shrines

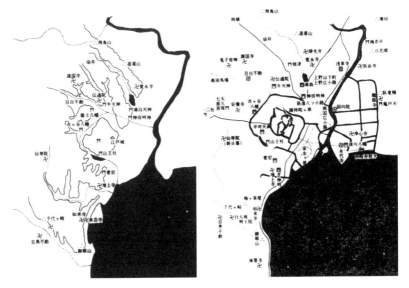

**Fig. 3-21** *Left:* The topography and temples and shrines of Edo (left); *Right:* The sights of Edo. From Masahiro Tanaka, *Nihon no kōen*.

in the castle towns of Japan.[27]
1. A temple district created on the periphery of a castle town where such institutions were grouped together (e.g., Takada, Tokushima, Yonezawa, Akita)
2. A continuous series of sites arranged on the outer edge of the castle (e.g., Matsuyama, Hirosaki, Uwajima, Fukushima)
3. Dispersed sites arranged at vital points inside the castle town (e.g., Takaoka, Ueda, Sendai)
4. Sites arranged at important points on or at entrances to roads in the castle town (e.g., Aizu Wakamatsu, Himeji, Kumamoto)

The planning of castle towns was characterized by clear zoning, and the above list of types of location makes it obvious that temples and shrines were intentionally arranged on the edge of town. This arrangement was not just for defensive purposes but seems to have signified an attempt to clarify the urban area visually and prevent uncontrolled sprawl. In Edo, while major temples and shrines were also located above bluffs, districts for groups of other temples and shrines were created on level land in Uchi-Kanda, Ryōgoku and Hatchōbori. These temples and shrines were transferred to the Asakusa district after the Great Meireki Fire (in the middle of the seventeenth century), and according to a *kiriezu* of Edo from around the Kanbun era (late seventeenth century), temples and shrines were arranged to encircle Edo from Ueno and Asakusa to Honjo. In this way, temples and shrines were continually moved outside by order of the shogunate as the city expanded; a belt of temples and shrines was arranged on level land that was open in the northeast direction to help define the outer edges of the city.

Temples and shrines were built not just above bluffs but in a specific relationship to microtopography. Even today, one finds an extraordinarily large number of temples and shrines in places where there are sharp breaks in the microtopography, that is, where contour lines on topographic drawings are concentrated. Temples or shrines and towns that develop along valleys often come in pairs; for example, Gokokuji and Otowadani, Hakusan Shrine and Sasugaya, Nezu Gongen Shrine and Nezudani, Sendagaya Hachiman Shrine and Sendagaya. The organization is the same in each case: the temple or shrine is built on high ground with its back to a hill, and the town forms on level land available in the valley in front of that temple or shrine. In Edo, areas were ordinarily named with reference to Edo Castle, but in the case of Otowachō, areas were numbered, from one to nine, with reference to Gokokuji temple. The higher the number the further was the area from the temple and the lower its gradient. In this way, planning was determined in relationship to a temple or shrine and not much affected by the surrounding environment. Temples and shrines are the cores of towns, but in Japan they were most often not located in the middle of valleys and did not become the center of town development. Furthermore, as can often be seen in local villages, the village shrine or the grove dedicated to the tutelary *kami* is not located in the center of the village, though it plays a central role in the community. Microtopography, that is, the fact that such a shrine has its back to a mountain or hill, probably accounts for the displacement of the shrine from the literal center of the community.

## SACRED MOUNTAINS

As can be seen in the Fuji cult of the Edo period, mountains have long been objects of worship. As with tumuli and mounds, the veneration of mountains is said to have begun in ancestor worship. According to Haruki Kageyama[28], the veneration of mountains began as primitive ritual in a clan society. It was believed that the spirits of clan ancestors who concealed themselves for a period of time in the mountains became *kami* and rose to heaven from mountain peaks and that guardian spirits attached themselves to mountains. The belief that mountains possessed a soul led gradually to a view of mountains as mysterious and sanctified; eventually mountains became sacred, forbidden areas. As a result, a nature-oriented form of religion developed in which mountains themselves appeared to be the object of worship. Tumuli rituals celebrating the spirits of ancestors that had been thus deified subsequently merged with agrarian rituals celebrating the natural spirits assuring bountiful harvests, thus creating the framework of Japanese Shinto, according to Kageyama. A shrine at the foot of a mountain is dedicated to the sacred mountain in the background and an inner shrine (*oku no miya*) stands on the peak of the mountain. However, in more primitive form, as in Miwa Shrine in Nara, there is no inner shrine in architecturalized form; instead, on the thickly wooded mountain peak there is only a rock called an *iwakura* to which the *kami* attaches itself. The *iwakura* is not always a rock but sometimes a waterfall, spring or large tree. In each case, a natural object that has seen careful, minimal human intervention has been selected. The *kami* who attaches itself to the *iwakura* moves in accordance to a seasonal, agrarian cycle, and the object of worship too moves accordingly. The prototype for this can be found in festivals celebrated all over the country. In its most common form, the *kami* is received by the *iwakura* (or *yama no miya*, the mountain shrine) on the mountain peak in spring, the most important season for agriculture, and then brought down to a specific place for a certain period. That place is generally at the base of the mountain on the outskirts of the village—the place

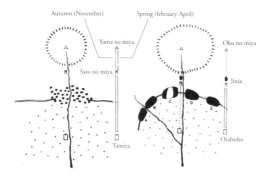

**Fig. 3-22** Primitive organization (left); organization of Shinto shrine (right). From Yūichirō Kōjiro, *Nihon no komyūnitii*.

(*sato no miya*) from which the mountain is usually worshiped. The *kami* may then be moved to the place where farming is actually performed (the *tamiya*, or rice field shrine). The *kami* that has been moved to a rice field will remain there until the autumn harvest and assure bountiful crops. When the harvest has been safely completed, the *kami* is thanked and sent back to the original mountain, thus completing the series of festivals. Since the object of worship is not in the center of the community, the *kami* is carried in a portable shrine, and a festival is a ceremony for that act of transfer. As religious forms eventually developed, the three sacred areas—*yama no miya*, *sato no miya*, *tamiya*—became the inner shrine (*oku no miya*), shrine (*jinja*) and the place where a portable shrine is positioned (*otabisho*)[29] (Fig. 3-22). The shrine that stands on high ground with its back to a hill on the outskirts of town is an architecturalized form of the *sato no miya*, and the thickly wooded hill behind it is meant to suggest the sacred mountain and inner shrine.

## WOODS (*MORI*) AND BULGE (*MORI*)

Mountains that are considered sacred and worshiped have a number of things in common. They are mountains that are close to town, visible from surrounding mountains, covered with woods and beautiful in appearance. Moreover, they must bulge up, like a bowl turned upside down, and be of a congenial size. Such mountains can be seen in villages all over Japan. Tumuli (*kofun*) are the first things that come to mind when considering such forms. They are mountains for *kami* constructed inside cities. In the sense that they are elevated urban symbols, they are said to be Japan's first tower-like constructs. The fact that Zōjōji in Shiba, an important datum in the planning of Edo, was built on the Maruyama Tumulus is symbolic. Though they are considered to have been symbols of power, tumuli were not made very high;

in time, overgrown with trees, they came to resemble natural hills. Tumuli do not rise toward heaven but instead hug the earth so that the qi in the ground is not dispersed into the air; they await the fermentation of the power of place.

Not just tumuli but all bulging landforms covered with trees, even a small hill in a corner of an estate, seem to have received special treatment. A heap of cut rice plants was called *nio* or *suzu*. *Nio* was considered a birthing hut, a sacred place where the soul of rice lay hidden until spring. That led to the practice of building a small shrine (*hokora*) dedicated to the *kami* of the estate including the *kami* of agriculture on a bulge in the ground in a corner of an estate. A bulging landform was regarded as a place where the spirit of the land lay hidden, and many mysterious legends survive concerning such places. The greenery covering the rise indicated a profundity of nature appropriate for concealment, and carelessly meddling with such a place was said to bring a curse[30]; a small shrine was built, and such places have often remained undeveloped even in modern times.

The idea of the guardian grove is said to have begun with the custom of gathering and worshiping the native *kami* in one place on the occasion of the development of a village for fear that encroachment on the land might bring a curse. Moreover, whenever a man-made reservoir was built, an island where the *kami* of the land or river could dwell was created in the middle of the water.[31] Thus, when human intervention on the land was necessary for purposes of development, a grove on an artificial mound was created to assure that the *kami* of the land had a place to reside. When nature had to be altered for human existence, certain protocols had to be observed. Nature was not to be completely eradicated; a part of it was to be captured and kept alive. Nature in such cases was not closely administered as with nature in a park; it was carefully tended but had to be highly natural nevertheless. Such places were parts of our immediate environment until not very long ago. Takeo Okuno[32] has explained as follows with respect to *harappa* (open field), which was integral to the primary landscape of Tokyo. "One realizes various things when calling to mind the appearance of *harappa*. There was generally a small shrine, stone Buddha or stone monument in a corner of the *harappa*. There was also a shrine-like structure, perhaps an Inari shrine, with a small *torii*. Old stone monuments or images dedicated to Kōshin shōmen, Taishakuten (one of the six heavenly beings, or Tembu, that are guardians of Buddhist law) or *dōsojin* might be gathered in one place; there were other places where a Jizō or a Batō Kannon (Hayagriva, a horse-headed deity of Hinduism and Buddhism) might stand by itself. There were a few old graves; there was also a south-facing slope called *hake* where water welled up from a spring. There would be a dense growth of large, ancient-looking trees such as zelkova or gingko. There was a *harappa* with manmade hills called Moto-Fuji (original-Fuji) or Shin-Fuji (new-Fuji) with origins in the Fuji cult, and another with a tumulus called Ōtsukayama." *Harappa* overgrown with grass was a highly natural *harappa*; it was not simply an empty urban lot but a sacred, taboo space that people could not easily alter. In that sense, an overgrown *harappa* had something in common with sacred mountains.

## THE *OKU* OF A VILLAGE

Sacred mountains greatly influence the way villages are organized. Examples in Japan of an entire mountain being developed to create a village are rare. As has been previously explained in connection with the types of microtopographic form, villages are generally embraced in the depths of mountains or stand between mountains (Fig. 3-23).

Yūichirō Kōjiro studied the three sacred places—*oku no miya*, *jinja* and *otabisho*—in actual villages and abstracted a prototypical organization for villages in Japan (Fig. 3-24).

"The archetypal Japanese community, in schematic form, was organized on a religious axis that went from the center of the community toward the sacred mountain; the axis passed through the shrine (*jinja*) at the foot of the mountain and, by way of the inner shrine (*oku no miya*) at the peak of the mountain, reached heaven. In the opposite direction, this axis descended from the center of the community to the riverbank and cultivated land where it was joined to the earth, the place of produce, by the *otabisho*. If one took the main road along the socioeconomic axis from the center of the community, one eventually reached the boundary between this community and the next." Thus he sees the basic form of the community as being organized along a

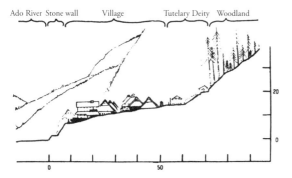

**Fig. 3-23** Sectional drawing of Machii Village. From Atsushi Ueda, *Ningen no tochi*.

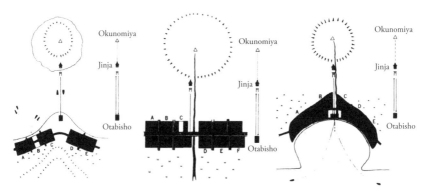

Fig. 3-24 Configuration of Japanese communities. From Yūichirō Kōjiro, *Nihon no komyūnitii.*

socioeconomic axis defined by the main road linking the houses together and a religious axis intersecting it.

The intersection of the two axes was certainly a center but not *the* central place in the community. By contrast, in many villages in the West the church stands on top of a small mountain and forms the very clearly visual center of the community. Houses stand all around the church; the entire mountain has been developed by the village. The church is truly located at the center of the village and can be distinguished from the houses by its sheer volume as well. The square in front of the church provides space to appreciate the structure's imposing volume, and the main streets converge on this square. The church tower makes apparent the territory of the community and integrates the houses; it is the center of the village in plan and forms the highest point on the mountain in elevation.

In Japan, the shrine located on the outskirts of town was a nucleus in the organization of the village. Just as the *okuzashiki*, the space of highest status, was located in a corner of the house in plan, the important place in the community was located in an out-of-the-way area. Though the shrine was as important a presence to its community as the church, it was located on the periphery of the village, and the inner shrine (*oku no miya*) was at an even greater remove. The inner shrine was buried in the woods of the mountain and not visible from the village. The three sacred places—*yama no miya, sato no miya* and *tamiya*—were quite inconspicuous in everyday life but their presence was felt on ceremonial occasions, that is, during festivals or at times of seasonal transition. On the approach to a shrine, *torii*, trees, lanterns, stone guardian dogs and straw festoons marked a series of boundaries and created a sense of depth much greater than the distance actually traveled. The repeated layering of such symbolic boundaries in a limited space gradually heightened the sacred character of the place. However, to understand this hierarchy, it was necessary to travel on a predetermined route and to experience these boundaries one after the next; such meanings were lost if one departed from that course. Unlike going to church by choosing any of the streets leading to the square in front of the church, going to a shrine by taking a short-cut that took one to the back of the precinct would have made the act of going to worship meaningless. Certain rules of behavior had to be observed in reaching that spot, and, moreover, the ultimate focal point always disappeared into the depths of nature. Nature that was unmanaged was a part of an unfathomable darkness, and unlimited depth was sensed in a space of dense growth. That is what led to the creation of a form of spatial organization based on a hierarchically arranged route that drew one into its depths.

NOTES

1. A "dissected valley" (*kaisekikoku*) is a valley formed through erosion by, for example, a river or glacier. The term is used in categorizing valleys by causation.
2. Microtopography (*bichikei*) is used here to mean landforms observable to the naked eye but not adequately expressed in ordinary topographic drawings. In geography, a distinction is made between macrotopography, intermediate-scale topography, small-scale topography and microtopography, but in this essay I have chosen to use microtopography in a broader sense to suggest the scale of topography of our immediate environment, one that is readily responsive to human intervention in the development of communities.
3. Ishige Naomichi ed., *Kankyō to bunka—jinruigakuteki kōsatsu* (Environment and Culture: Anthropological Studies) (Tokyo: Nihon Hōsō Shuppan Kyōkai, 1978).
4. Higuchi Tadahiko, *Keikan no kōzō* (Tokyo: Gihōdō, 1975), trans. Charles Terry, *The Visual and Spatial Structure of Landscapes* (Cambridge: MIT Press, 1988).
5. Four sacred beasts—imaginary animals—ruling the four directions of heaven in Chinese mythology. They correspond to the four directions, the four seasons and the four colors.
6. Feng shui was used in ancient China to judge the felicity of locations for cities, residences and other buildings, and tombs. It was based on the idea that the flow of *qi* can be controlled through the location of things.
7. In Chinese philosophy, qi is generally considered to be invisible and in flux and to have an effect on things. As with the Latin *spiritus*, it is related to the idea of breath understood as vitality or spirituality.
8. In Japanese, when two words are joined to form a compound word, the initial consonant of the second word is "voiced;" hence, *kami* combined with *yashiki* (estate) produces *yashikigami*, and *saka* combined with Dango (dumpling)—see Section 2—produces Dangozaka.
9. Motoaki Tawara, "Kono ningenteki na chizu" (This Human Map), *Nihon no kochizu* (Tokyo: Kōdansha, 1976).
10. Fukutarō Okui, *Toshi no seishin* (The Spirit of Cities) (Tokyo: Nihon Hōsō Shuppankai, 1975).
11. The distinctive microtopography of Edo results from the erosion of a plateau that was 60 *shaku* above sea level by water systems, creating a number of peninsula-shaped plateaus around Edo Bay. That is why it can be indicated by a single 60-shaku contour line.
12. So-called *banchō* districts reserved for high-ranking samurai and located near the castle (the most famous being the one in what is now Chiyoda-ku, Tokyo) and *monzen-machi* (towns that developed in the vicinity of powerful temples and shrines) were exceptions; there, lots were numbered.
13. Eichi Yokoseki, *Edo no saka Tōkyō no saka* (The Hills of Edo, the Hills of Tokyo) (Tokyo: Arimine Shoten, 1976).
14. Bunkyō Kuyakusho, ed., *Bunkyōkushi 1* (History of Bunkyo Ward 1) (Tokyo: Bunkyo Kuyakusho, 1981).

15. *Man'yōshū*, the oldest existing collection of Japanese poetry, edited from the late seventh century to the late eighth century.
16. Noboru Kawazoe, *Toshi to bunmei* (City and Civilization) (Tokyo: Sekkeisha, 1966).
17. Toshio Kitami, *Nihon minzokugaku no shiten* (The Point of View of Japanese Folklore).
18. *Kekkai*, originally a Buddhist term signifying the separation of sacred and profane areas, came to mean spatial boundaries in a broad sense.
19. Kan'hasshū was the name given the Kantō region in the Edo period.
20. Shinjirō Kirishiki, *Edo.Tōkyo no toshishi oyobi toshikeikakushi kenkyū* (Studies in the Urban and City Planning History of Edo/Tokyo) (Tokyo: Tōkyō Toritsu Daigaku Toshi Kenkyū Soshiki Iinkai, 1971).
21. The axis of grids on plateaus was generally in accord with the orientation of the topography, and mountains and hills were possibly used as survey data in making finer adjustments.
22. Kirishiki, *Edo. Tokyo*
23. *Tokomise* was a small, movable store or stall.
24. In the planning of Edo, *fushingata* of the shogunate were officials responsible for public works and *sakujigata* were officials responsible for buildings; garden projects were under the responsibility of *fushingata*.
25. Masahiro Tanaka, *Nihon no kōen* (Parks of Japan) (Tokyo: Kajima Institute Publishing Co., Ltd, 1974). "Tiles" refer to roof or ridge tiles.
26. Tatsuo Hagiwara, *Edo no Inari* (Inari of Edo).
27. Enjirō Yajima, *Nihon toshi hattatsushi* (History of the Development of Japanese Cities) (Tokyo: Hōbunkan Shuppan, 1969.)
28. Haruki Kageyama, *Shintaisan* (Sacred Mountains) (Tokyo: Gakuseisha, 1971).
29. *Otabisho*, a place where the portable shrine for carrying the *kami* during a festival is placed for a brief time or overnight.
30. A curse (*tatari*) was originally a demonstration of the enormous power of a *kami* but later came to imply punishment wrought by the *kami*.
31. Tembi Kanai, *Shitsugen saishi* (Moor Rituals) (Tokyo: Hōsei Daigaku Shuppankai, 1975).
32. Takeo Okuno, *Bungaku ni okeru genfūkei* (Primary Landscapes in Literature) (Tokyo: Shūeisha, 1972).

CHAPTER 4

# THE EXTERNAL LAYERS OF STREETS

*Hidetoshi Ohno*

## INTRODUCTION

Landscapes in the city are varied. Even if we limit ourselves to landscapes seen from that most commonplace of viewpoints, the street, and, what is more, to streetscapes in residential areas, the subject of this essay, we encounter enormous variety. In Tokyo, for example, streetscapes in suburban residential districts[1] in the southwestern part of the metropolis are different from those in the old *shitamachi* districts of Edo to the east, and even within the same district, the landscapes we find on back streets are entirely different from those we find on thoroughfares.

Streetscapes are familiar sights in the urban environment and for that reason have a direct bearing on our characterization or assessment of an area or place. Furthermore, streetscapes are part of a shared experience; some become representative or symbolic of an area or place and thus part of the image of the city as a whole. Streetscapes are involved in all our experiences, ideas and images concerning the city. That is because we do not use the material environment simply to utilitarian ends; we situate ourselves in the world, temporally and spatially, through the mediation of landscapes.

If we are to analyze streetscapes, we must first identify the elements from which they are composed. Figure 4-1 shows a streetscape in an Italian city. There, the compositional elements are a street and buildings standing on that street. In the Tokyo streetscape shown in Figure 4-2, the elements are more than a street and the buildings on it. Trees rise behind a fence and potted plants are arranged under eaves. Laundry has been hung out to dry on a second floor. The streetscape is formed from all these things in combination with the buildings. In a more upscale residential area of the kind referred to as *oyashiki-machi*,[2] on the other hand, the buildings, standing behind property walls and trees, are often invisible from the street. There is no appropriate term in the vocabulary of architectural or urban studies for such a combination of streetside elements. For the purpose of analyzing and discussing Japanese streetscapes, therefore, I will henceforth call that combination the "external layers" (*hyōsō*). We may then say that the compositional elements of a streetscape are streets and external layers. The abovementioned Italian streetscape can be regarded as one having a special form of external layers composed only of the facades of buildings.

The forms external layers take are more than the result of separate and distinct conditions such as the size of the property or the taste of the residents; they are also a matter of culture.

External layers are the interface between buildings and the city. Thus, an

Fig. 4-1 An Italian streetscape: San Gimignano. The street space is bounded by the walls of buildings.

Fig. 4-2 A Tokyo streetscape: Toranomon, Minato-ku. The street space is bounded by not just the walls of buildings but by hedges and plants; they combine to participate in the streetscape.

investigation of external layers can be said to be the nexus between urban studies and architectural studies. In this essay, we shall consider the reasons for formal differences in external layers from two perspectives. The first is that external layers are the boundary between two kinds of urban territory, the private domain and the public domain (i.e., the street). The second is that forms of external layers are, like architectural and urban forms, products of particular views of nature and spatial concepts.

The external layers of a city, unless they are part of, say, a planned housing project, are not designed by experts. Instead, they are the cumulative result of building activities by many different people. Nevertheless, a number of types of external layers are identifiable in residential areas in Tokyo, and distinctively Japanese characteristics can be determined if those types are compared with their European counterparts. This suggests that within a certain group a shared view of external layers exists, and that there is a tacit or unconscious agreement regarding that view. I hope to provide some idea of that view of external layers and the underlying urban or architectural perspective through an analysis of the forms they take.

I will deal mainly with the external layers of residential areas because houses are the basic elements of cities. Houses have developed into a building type with only a residential function in the modern city. In premodern cities, however, houses could also be workplaces, offices or shops, and large houses served as palaces. Palaces evolved in later eras into art museums and hotels. Houses thus constitute an important and basic type of building from which cities are composed. Moreover, houses are the domain of the smallest and most universal group from which society is organized, the family, and, because of the conservative nature of habitation, are deeply rooted in the culture to which they belong.

## 1. THE EXTERNAL LAYERS OF RESIDENTIAL AREAS

### FOUR TYPES OF EXTERNAL LAYERS IN RESIDENTIAL AREAS

Several types of external layers exist in residential areas. I will attempt to categorize those types both by the way external layers are organized in section in a direction perpendicular to the axis of the street and the way houses[3] are arranged along the street.

First, buildings are not often constructed right up to the street in residential areas in Japan. If we look at the sectional organization, we see that the distance from the street to the main building is a factor determining the character of the external layers. The condition of the street boundary is another factor. For example, a high fence may conceal what is beyond it from

CHAPTER 4  THE EXTERNAL LAYERS OF STREETS   107

Fig. 4-3 *Left:* External layers with a wall-like secondary plane: Toranomon, Minato-ku, Tokyo.
Fig. 4-4 *Right:* External layers with a porous secondary plane: Miyamae, Suginami-ku, Tokyo.

view (Fig. 4-3), while a low hedge or a wire screen may allow views of, say, children playing in the yard (Fig. 4-4). If, for the sake of convenience, we call the plane that defines the boundary of the house with the street the first layer, and the wall of the main building on the streetside the second layer, then two factors—the distance between the first layer and the second layer and the properties of the first layer—can be said to determine to a large extent the character of the external layers. There is to each of these two factors a threshold that differentiates the character of the external layers. With respect to the distance between the two layers, the type of situation in which the main building is constructed up to the street is at one pole and the type of situation in which the main building is located at a far distance from the street is at the

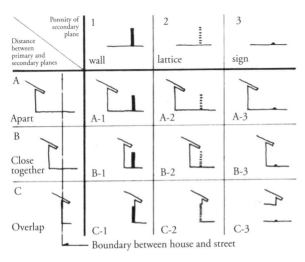

Fig. 4-5 Types of external layers with different sectional organizations.

other; the type of situation in which there is sufficient distance between the two layers to park an automobile is halfway between those two extremes. With respect to the properties of the first layer, the type of situation in which what is beyond that layer is totally hidden, as behind a solid wall, is at one pole, and the type of situation in which there is only a token acknowledgment of the boundary between the house and the street (such as a difference in paving, a low stone border or a hedge) is at the other; the type of semi-porous situation in which, for example, lattices offer a glimpse of what is beyond the barrier is halfway between those two extremes.

Three stages have been identified for each of two factors; this table (Fig. 4-5) shows nine possible scenarios. With this table in mind, let us examine actual residential areas in Tokyo.[4] Type A-1, where the first layer is a solid wall and the second layer is at a distance from the first, represents the external layers of an upscale house (*oyashiki*) (Fig. 4-6). There is a garden beyond the wall, and the tops of trees rise above the wall, at times spreading their branches over the street. The roof of the main building can be glimpsed beyond those branches. This type of external layers, made up of few elements, tends to be relatively monotonous and restrained. At night, lights from inside the house are not visible, and the street is quiet.

Type A-2, where the first layer is semi-porous, is not to be found in the heart of Tokyo; the external layers of a farmhouse situated beyond the residential suburbs in an area near Tachikawa along Itsukaichi Highway[5] are of this type (Fig. 4-7). Here, we glimpse the main building beyond a low hedge and the zelkova trees lining the road. Type A-3, where the second layer is at a distance from the first, and only a hint exists of that first layer, is also not to be found in the Tokyo suburbs. The external layers of a farmhouse in the village of Ōuchi, Fukushima prefecture (Fig. 4-8), are of this type. Examples of this type are rare in Japan but commonplace in the suburbs of North America, where neat wooden houses standing at the far end of manicured lawns and at

Fig. 4-6 An example of A-1: Takanawa, Minato-ku, Tokyo.

a slight remove from neighboring houses on either side are a familiar sight. At times, only a mailbox by the side of the road indicates that a lawn is private property.

When the distance between the first and second layers closes, the second

Fig. 4-7 *Top left:* An example of A-2: suburb of Tachikawa, Tokyo.
Fig. 4-8 *Center left:* An example of A-3: Ōuchi, Fukushima prefecture (photographer: Tsuneaki Nakano).
Fig. 4-9 *Bottom left:* An example of B-2: Shōan, Suginami-ku, Tokyo.
Fig. 4-10 *Top right:* An example of B-3: Negishi, Taitō-ku, Tokyo.
Fig. 4-11 *Bottom right:* Potted plants in an alley: Hongō, Bunkyō-ku, Tokyo.

layer also participates in the external layers and the number of elements composing the external layers increases.

Even when the first layer is a wall (Type B-1), elements such as windows, eaves, shutter boxes (*tobukuro*), and balcony handrails of the main building can be seen from the street and create lively external layers if the main building is two-storied (Fig. 4-3). When the first layer is simply a hedge or a wall with many openings (Type B-2), the second layer participates to an even greater extent in the external layers (Fig. 4-9). These are the two types of external layers most frequently and widely found among low-rise detached houses in Tokyo.

When the porosity of the first layer increases further and simply suggests the presence of a boundary between the property and the street (Type B-3), the second layer assumes the main role in the external layers, and the external layers become less complex once more (Fig. 4-10). This is the townscape that traditional townhouses (*machiya*) once presented. (*Machiya* were the houses of *chōnin*, that is, merchants and artisans, as opposed to members of the ruling samurai or warrior class.) *Shimotaya*-style townhouses[6] are a type of *machiya* rarely found in Tokyo today but still seen in cities throughout Japan (Fig. 4-13). The main buildings of such *machiya* are sometimes constructed, side by side, up to the street; however, more commonly the building wall is set back slightly from the boundary between the property and the street indicated by the curb or a watercourse and creates, together with the eaves overhead, an ambiguous domain between the street and the main building. Fence-like devices such as *inuyarai* and *komayose* found in traditional streetscapes have the psychological effect of clarifying the boundary of the property and keeping passersby at a distance from the main building. Much the same can be said of the potted plants arranged outside rowhouses on back streets (*ura-nagaya*) today. Even in Tokyo, potted plants are often found in alleys (*roji*)[7] that extend from thoroughfares into the inner recesses of blocks in the *shitamachi* districts first developed in the Edo period. Small houses are built cheek by jowl along such lanes. These *ura-nagaya*[8] often have potted plants and bonsai arranged in front of them that not only add color to a crowded urban area but effectively transform the alley into a public place. *Machiya* and *ura-nagaya* have external layers with the same sectional organization, but the external layers are used in different ways. Houses facing alleys—that is, *ura-nagaya*—have lively external layers that include not only potted plants but objects such as bicycles, household items and platforms at the second-floor level used for hanging laundry. By comparison, the external layers of *machiya* are more ordered.

Types of external layers where the main building fronts directly on the street, shown in the bottom row of Figure 4-5, are not common in the residential areas of Tokyo or for that matter, Japan. As has already been explained, the external layers of *machiya* and *ura-nagaya*, though often said to be in direct contact with the street, are on closer examination found to be separated from the street by a space, however slight. The courtyard type of house widely found in Arab cultures has external layers of Type C-1. No openings are needed on the streetside wall because the courtyard provides ventilation and light. Few houses of the courtyard type have been built in Japan.[9]

Type C-2, where the first layer is semi-porous, and Type C-3, where the inside is highly visible, are external layers suitable for houses that also serve as shops needing to display wares to passersby. They do not suit houses that are only residential in function where privacy is a concern.

This examination shows that external layers in Tokyo residential areas are mainly of Types A-1, B-1, B-2 and B-3. These can be arranged into three groups: group 1, where the second layer is the main constituent of the external layers (B-3); group 2, where the first layer is the main constituent of the external layers (A-1); and group 3, where both the first and second layers participate in the external layers (B-1, B-2). The character of the external layers is determined by not just the sectional organization but also the way houses are arranged along the street. The *oyashiki* type (with a sectional organization of Type A-1) and the *machiya* type (with a sectional organization of Type B-3) differ in that the wall in the case of the former is a property wall and in the case of the latter is the wall of the main building, but both types abut the walls of adjacent houses and creates a continuous plane (Figs. 4-12,13). The *ura-nagaya*, which has a sectional organization similar to that of the *machiya*, also abuts the houses next door, but houses of this type are irregularly arranged. Since the plane of the main building may be pushed forward or set back, houses of this type do not create a continuous wall plane (Fig. 4-14). There is no apparent datum for the arrangement of a house as there is with the *machiya* type or the *oyashiki* type. The boundary is made even more ambiguous by potted plants and household items in the alley. An outsider who wanders into an alley feels as if he or she has accidentally intruded into another person's private space. One reason for this is the fact that objects such as laundry and household items that are concealed in the back of the house in the case of the *oyashiki* type or the *machiya* type are here exposed in the front. That is, it feels as if one has wandered into someone else's backyard. The other reason is the closed-off quality of an alley. An alley is narrow in width, and moreover often a cul-de-sac or L-shaped (Fig. 4-16); at times, a building covers the entrance to the lane, creating a tunnel-like effect. "Public distance" (3.6 meters or more) as defined by Edward T. Hall,[10] who studied the relationship between

Fig. 4-12 *Top left:* External layers of the *oyashiki* type: Honchō, Nakano-ku, Tokyo.
Fig. 4-13 *Center left:* External layers of the *machiya* type: Ōmihachiman, Shiga prefecture.
Fig. 4-14 *Bottom left:* External layers of the suburban house type: suburb of Kashiwa, Chiba prefecture.
Fig. 4-15 *Right:* External layers of the *ura-nagaya* type: Hongō, Bunkyō-ku, Tokyo.

distance and human communication, is difficult to achieve in an alley. That is, unable to stay adequately outside the sphere of involvement, one finds oneself within the "social distance" (1.2–3.6 meters) of residents living behind those external layers.

To differentiate between external layers of the *machiya* type and the

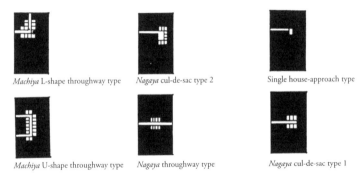

Fig. 4-16 Alley patterns. From Noboru Shimamura, Yukio Suzuka, et al., *Kyō no machiya*.

*oyashiki* type on the one hand and external layers of the *ura-nagaya* type on the other, let us call the former "epidermal" and the latter "endodermal." The epidermis encloses the organism and is smooth and defensive in character. The endodermis, on the other hand, as can be seen in the inner walls of digestive organs, has many folds or creases and is open and receptive.

A mixture of the first and second layers creates in the streetscape external layers that are constantly interrupted and somewhat irregular. This type has continuous streetside property walls or hedges, as with the *oyashiki* type, but such elements are often low in height. We often find elements that are highly porous such as louvers and wire mesh. As a consequence, the exterior appearance of the main building, which forms the second layer also participates in the external layers. A main building of this type does not abut the houses next door as with *machiya* or *ura-nagaya*; like a main building of the *oyashiki* type, each unit is independent. As a result, external layers of this type are, like a comb, connected at the bottom but discontinuous at the top (Fig. 4-15). We shall call it the "suburban house type" since it is typically found in the suburbs.

As with the *ura-nagaya* type, external layers of the suburban house type may provide glimpses of private life within the house. At night, lights from windows illuminate the street; its streetscape thus differs from the dark streetscape of the *oyashiki* type. However, unlike a neighborhood of the *ura-nagaya* type, a street of the suburban house type is free of household items.

As we have seen, four types of external layers can be identified for houses in Tokyo, namely, the *oyashiki* type, the *machiya* type,[11] the *ura-nagaya* type and the suburban house type.

Let us try locating these four types of external layers along two axes to identify their interrelationships (Fig. 4-17). Though they differ in the

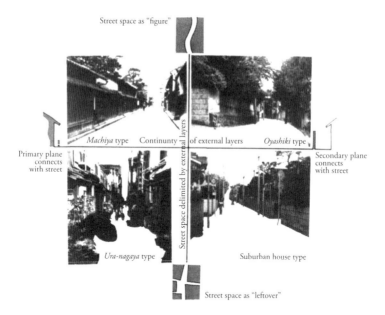

Fig. 4-17 Formal structure of external layers.

spaciousness of their sites, the *oyashiki* type and the suburban house type use the space of the site in similar ways. In both cases the main building is surrounded by a garden and at a remove from the main buildings of the houses next door. The first layer—that is, property walls and hedges surrounding the main building—abuts the first layers of the houses next door. By contrast, in both the *machiya* type and the *ura-nagaya* type, the second layer abuts the second layers of the houses next door, and the first layer is tenuous and merely suggested.

The formal character of the street is influenced by the form of the external layers delimiting the street space. External layers of the *oyashiki* type and the *machiya* type are continuous and epidermal in character; as a result, the street space reads as "figure" and emerges as a clear object. By contrast, external layers of the suburban house type and the *nagaya* type lack a clear order and are endodermal in character; in neither case does the street space possess coherence. The street space appears to have been left over by external layers. That is, the four types of external layers in residential areas can be organized along a horizontal axis indicating which layer abuts the houses next door and a vertical axis indicating the character of the street space created by the external layers.

The external layers are the boundary between a house and the street.

CHAPTER 4  THE EXTERNAL LAYERS OF STREETS   115

Fig. 4-18 *Oyashiki* in Edo. *View of Kasumigaseki* by Ando Hiroshige (1797–1858), woodblock print.

However, unlike the boundary between a house and the house next door, they are located between the public domain (the street) and the private domain (the interior of the house) and are a spatial device that serves as a valve controlling social relationships with, for example, acquaintances and guests. Figure 4-17 also expresses the degree of communication between the inside of a house and the street. The relationship between inside and outside is weakest in the *oyashiki* type at the top left corner of the figure and becomes closer the further to the bottom right corner we go; the relationship is closest in the *ura-nagaya* type. There is little pedestrian traffic in a neighborhood of *oyashiki* type houses; garden trees extend their branches over quiet streets. Passersby are given few hints of what is happening beyond property walls. On the other hand, as soon as one enters an alley of *ura-nagaya* type houses, the windows of houses are right in front of one; if, as in summer, they are open, one can see people relaxing in living rooms. As has already been explained, the narrow width of an alley and the smallness of houses ensnare everyone within what Hall calls "social distance." One sees friends conversing across a window sill. At night, lights from houses spill out onto the alley, and one can hear voices and the sounds of television sets.

There were three forms of urban houses in Edo-period Japan: *buke jūtaku* (houses for samurai), *machiya* and *nagaya*.[12] Three types of external layers in residential areas today are closely related in composition to the external layers of those three forms of houses (Figs. 4-18, 19, 20). A fourth, the suburban house type of external layers, can be understood to be the external layers of *buke jūtaku* in miniature.[13]

The above has been a discussion of the external layers of the contemporary city. Next, we shall consider the spatial structure of the external layers of the three house forms of the Edo period and their function.

## *BUKE JŪTAKU* AND *MACHIYA*

As has already been explained, the external layers of a house are located between the public domain and the private domain and serve to control social relationships with, for example, acquaintances and guests. Let us consider this in more detail.

In *buke jūtaku*, the main building is surrounded by outdoor spaces such as gardens and service yards, which are in turn surrounded by a property wall; in a grand residence of this kind, *nagaya* where retainers and servants live are attached to the inside of the property wall. Thus, the elements comprising

**Fig. 4-21** Example of the main residence of a *hatamoto* (a high-ranking vassal of the shogun), Nabeshima Takuminokami Naotaka. From Akira Naitō, *Edo to Edojō*.

the external layers are a closed wall, a garden or gardens, and at times the roof or wall of the main building; the entrance is built into those elements. The entrance is composed of the gate (a break in the property wall), the forecourt (ordinarily walled off from other outdoor spaces), and the *genkan* (the ceremonial entrance to the main building) (Fig. 4-21).

This set of spaces—gate, forecourt and *genkan*—is widely found in entrances to *buke jūtaku*. As a document (Fig. 4-22) describing the way visitors to the Edo residence of the Kaga domain are to be welcomed or seen off indicates, it plays an important role in the reception of guests. As the document makes clear, various critical points—starting with the *amaochi* (a channel or stone placed where rain falls from the eaves) of the gate and ending

| Visitor's status | To welcome or to send off | Person doing the welcoming or the sending off | Place for, and the circumstances surrounding the welcome or send off | |
|---|---|---|---|---|
| Same rank as the master of the house | Welcome<br>Send off | Samurai in charge of the entrance<br>{ Master of the house<br>{ Samurai in charge of the entrance | *Amaochi* of the gate<br>Boarded area of the *genkan*<br>*Amaochi* of the gate | 2 persons, prostrate themselves side by side<br><br>2 persons, prostrate themselves side by side |
| Lieutenant general (*chūjō*) | Same in either case | Samurai in charge of the entrance | *Amaochi* of the gate | 2 persons, side by side |
| Major general (*shōshō*) | Same in either case | Samurai in charge of the entrance | Pebbled area (*Shirasu*) | 2 persons, side by side |
| Highest ranking official of the shogunate (*rōjū*) | Welcome<br>Send off | Samurai in charge of the entrance<br>{ Master of the house<br>{ Samurai in charge of the entrance | Boarded floor area<br>Wide veranda<br>Boarded floor area | 2 persons<br><br>2 persons |
| *Omote daimyō* of at least 100,000 *goku* | Same in either case | Samurai in charge of the entrance | *Boarded floor area* | 2 persons |
| *Daimyō* of at least 10,000 *goku* | Welcome<br>Send off | Samurai in charge of the entrance<br>Samurai in charge of the entrance | *Kagamiita*<br>*Kagamiita* | 2 persons<br>2 persons |

**Fig. 4-22** Ways of welcoming and sending off visitors to the Edo residence of the Kaga domain. From Kiyoshi Hirai, *Nihon no kinsei jūtaku*.

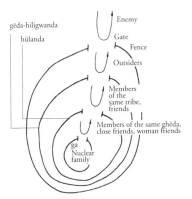

**Fig. 4-23** Hierarchy observed in receiving guests in dwellings of a Datooga tribe of Tanzania. From Naomichi Ishige, *Jūkyo kūkan no jinruigaku*.

at the panel board (*kagamiita*) of the ceiling over the *genkan*—were identified in the space stretching from the gate through the forecourt to the *genkan* for use on such occasions, depending on the status or precedence of the visitor (who could range from a personage of the same rank as the master of the house to a *daimyō* of at least 10,000 *koku*).[14] Just as meetings and receptions in the formal reception room—the *shoin* or *hiroma*—were ritualized in great detail, the way people were welcomed or seen off at the entrance was also ritualized. This shows that the entrance space was a place indispensable for the receiving of guests.

People today are apt to think of a reception room (*ōsetsuma* or *zashiki*)[15] when the subject of attending to visitors is raised, but if it includes all the ways in which people outside the family are dealt with inside the house, then it begins at the entrance to the house. Naomichi Ishige makes an interesting point regarding this question.[16] According to Ishige, behaviors engaged in within a house are common to all human societies. They are sleeping, resting, caring for the young, educating, cooking, eating, managing household items, attending to visitors and secluding the household from the outside world. Since animals in their lairs also sleep, rest, care for the young, educate and eat, "cooking, managing household items, attending to visitors and secluding the household from the outside world are the ways of use that differentiate human domestic spaces from animal lairs."[17] He points out that as far as attending to visitors is concerned, a four-stage hierarchy is observed in the dwellings of the Datooga tribe of Tanzania depending on the type of visitor. This spatial hierarchy is similar to the one found in the Edo residence of the Kaga domain. This kind of hierarchy for attending to visitors is an expression of the social relationship between the master of the house and the visitor. For example, if a person who has been received at the *genkan* is invited into the house, this means that a close relationship has been acknowledged. If a visitor is admitted into the family circle in the living room, it signifies that he or she is someone with whom the host feels at ease. To admit a visitor into the formal reception room (*zashiki*) indicates respect.

A house must have a spatial order—a number of critical points—to make such a staged approach in attending to visitors possible. The Hiroshi Hara Research Group at the University of Tokyo saw such critical points as "thresholds" and considered the organization of houses as a problem of the nature of thresholds arising from the interface between two domains (the family domain and the community domain).[18] Their survey of the Mediterranean region suggested a number of models of houses. Concerning one such model from Spain, the Petrés type, they explain as follows. "Even people outside the family are freely admitted into the foyer-like room

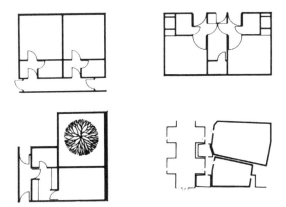

Fig. 4-24 Locks in action. From Serge Chermayeff and Christopher Alexander, *Community and Privacy.*

adjoining the street… However, they are never admitted beyond that room. This place into which others are not permitted to enter is precisely the domain of the closed family circle… We have chosen to call this foyer-like part of the house a 'combining threshold' (*ketsugōiki*). A combining threshold is a device that enables two or more domains to maintain mutual contact without interfering with one another.[19]

Sergei Chermayeff and Christopher Alexander developed a similar concept for architecture in general (Fig. 4-24).[20] "The idea of locks between different activities as a planning device is not new. In hospitals, contaminated general purpose areas are separated from sterile areas such as operating rooms by sterilization locks through which all persons and equipment must pass… In all these instances, however, the lock is virtually a passage: a secondary transition between two major zones. It is intended here that the lock eventually become as important as any other zone of activity."[21]

The space created by the gate, forecourt and *genkan* in *buke jūtaku* was clearly a device like the "combining threshold" or "lock." It was a compartmentalized space at the interface between two domains. It was where the mode of action was altered; traditionally visitors alighted from palanquins or horses, and accompanying servants were made to wait. A place where such attendants could sit was provided facing the forecourt. It was also the first checkpoint for protecting the family domain; there were places such as the *tōsaburai* and *okachimetsuke* just beyond the *genkan* and the *bansho* by the side of the gate where guards and keepers waited.

Such a spatial device is necessary because two contrary demands (seclusion and contact) are made on households. The family, as a group with its own internal order, has to be independent of society, but at the same time the family, as a part of society, has to maintain a connection with society for its continued existence. These two demands are most exigent for the *machiya*. The *machiya* is essentially a combined shop and house; its surface is stretched the full width of the site in an attempt to attract as many customers as possible. However, this means that family life is made vulnerable to various negative factors of high-density life such as fire in adjacent houses, visual intrusion and noise from adjacent houses and the street, and burglars at night. A distinctive spatial device was created on the front side of the *machiya*, where this contradiction is most apparent.

An earthen-floored passageway-and-entrance called *tōri-niwa* stretches from the front to the back on one side of a *machiya*. Along this space are arranged at a minimum one row, and in some instances many rows, of rooms. Of these rooms, the room facing the street is often called *mise* or *misenoma* (shop), whether the house is a shop or (as in *shimotaya*) no longer engages in commerce. The spaces along the street (including the *tōri-niwa*), which we shall call "mise spaces" are formally differentiated and separated from the spaces behind them. For example, in *machiya* such as those called *kagura-date* in Mikuni-machi (Fig. 4-25) and *hamaya-zukuri* in Wajima,[22] the ridge of the roof over the *mise* spaces is parallel to the street, while the roof over the central portion of the building is oriented in a direction perpendicular to the street. There are also many examples where *mise* spaces are accommodated in an entirely separate structure. In Edo, many shops were made of wattle-and-daub construction (*misegura*), while the living quarters in the back were made of ordinary wooden construction. In this way, the two areas were differentiated by structural method.

*Machiya* tended to be built on deep sites with narrow frontages; to enlarge a house, rooms were added in the back. A small courtyard (*tsubo-niwa*) was introduced to enable rooms in the middle of the house to access daylight. The *tsubo-niwa* was often located directly behind the *mise* spaces. There were also instances where a corridor or stairs were arranged behind the *mise* spaces. Furthermore, partitions arranged in the longitudinal direction in the *mise* spaces did not always align with those in the rooms in the back, as in the example of a *machiya* in Hikone (Fig. 4-26). A distinction was also made between the front and back areas of the *tōri-niwa*. Not only was there a door (*nakado*) between the two areas, the *mise* spaces had a one-story ceiling height while the space beyond the door had a double-height ceiling.[23] The *mise* spaces facing the street in the Yoshijima House in Takayama, Gifu prefecture, are distinguished by the same form of ceiling—one with exposed joists—even though one space is earthen-floored and the other has a raised tatami floor.[24]

As the above has shown, rooms in a *machiya* facing the street, that is, *mise*

CHAPTER 4   THE EXTERNAL LAYERS OF STREETS   121

Fig. 4-19 *Machiya* in Edo. Hiroshige, *Sketch of Nihonbashi Tōri Itchōme* by Ando Hiroshige, woodblock print.

Fig. 4-25 *Kaguradate*-style of *machiya*, Mikuni-machi, Sakai-gun, Fukui prefecture. From Atsushi Ueda and Atsuo Tsuchiya, eds., *Machiya–kyōdo kenkyū*.

1. Sedo
2. Zashiki
3. Oe
4. Mise
5. Kōshi
6. Shitomido

1. Storehouse
2. Hanare
3. Zashiki
4. Nakanoma
5. Denji
6. Latrine

Fig. 4-26 Non-alignment of partitions, Hikone, Shiga prefecture.. From Ueda and Tsuchiya, *Machiya–kyōdo kenkyū*.

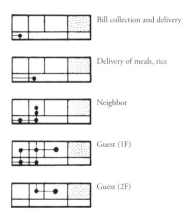

**Fig. 4-27** Hierarchy observed in receiving guests in a Kyoto *machiya*. Shimamura, Suzuka, *Kyō no machiya*.

spaces, tend to be clearly differentiated in form on the inside and the outside from the rooms behind them. The differentiation of the *mise* spaces is a reflection of the hierarchy observed in attending to visitors. As the name *mise* (shop) indicates, the spaces serve commercial functions, generally the display of merchandise, when the building in question is a shop. People are permitted into *mise* spaces; indeed, it is hoped that they will enter. However, the area past the door called *nakado* is family domain; outsiders are not allowed into it without permission (Fig. 4-27). A similar hierarchical organization in attending to visitors is found in *shimotaya*, which are used exclusively as dwellings. In *shimotaya*, a *mise* space is most often used as a room for receiving visitors (*ōsetsushitsu*).

*Mise* spaces might be likened to a "thick wall" arranged between the family domain and the public domain, that is, the street. *Mise* spaces in *shimotaya* are also used as spare rooms or storage, at times as private rooms for elderly persons or children. Rooms important to the family such as the family/dining room, the room for receiving formal guests (*zashiki*) and the sleeping room for the master and mistress of the house—what might be referred to as the *ie*, a term for "house" that also meant "a group ruled by the head of the house" until the end of World War II period—are located in the rear, protected from the noise of the city and the eyes of passersby by the "thick wall" of the *mise* spaces.

Though functioning as a "wall," *mise* spaces are not completely closed. Louvers are installed in front, promoting ventilation. Moreover, as is well known, louvers make it easy to see from the side that is darker to the other side but difficult to see in the opposite direction (Figs. 4-44, 51). During the day, passersby cannot see beyond the louvers even if the *shōji* screens behind those louvers are opened, but persons inside can easily see what is happening on the street.

"Before leaving, the master of a shop in Nishijin, Kyoto, is said to look outside from behind louvers and ascertain that no one in the neighborhood is around."[25]

"There were fewer family members now, and the *koshinoma* [literally, "a room with louvers"] became grandmother's room. Through the louvers she could see perfectly the state of the neighborhood, for example, whose son was gadding about and going out on the town at night."[26] Watching passersby is one of the pleasures of a streetside cafe, but that sort of urban amusement is impossible in a *buke jūtaku*. The louvers of *mise* spaces are removed during festivals such as the Takayama Festival in Gifu or the Gion Festival in Kyoto; red carpets are laid, screens are brought out, and decorations are arranged. On such occasions, *mise* spaces are directly connected to the public domain. *Mise* spaces are indeed a third domain where two different domains overlap. Being a place of change in mode of movement, *mise* spaces are equipped with shelves for clogs, umbrella stands and storage; bicycles, motorbikes and baby strollers are parked there; at times they become a garage for a *machiya* without a proper place for an automobile.

The gate-forecourt-*genkan* space in a *buke jūtaku* and the *mise* spaces in a *machiya* do not function simply to connect or disconnect the family domain from the public domain. They make possible a staged transition (in style of action or method of movement) from the outside to the inside. These spaces must be of sufficient size to serve such a function. If, unaware of the importance of such spatial devices in houses, people unnecessarily use spaces of status as an entrance, the family has no place to gather when a guest is being received; cold winds blow in the moment the front door is opened; and people at dinner are exposed to view. Life in the house becomes miserable.

Fig. 4-20 *Ura-nagaya* in Edo. From *Gōkan ehankiri kakushi no fumitsuki*.

## *URA-NAGAYA*

A single unit of *ura-nagaya* in an alley is in a sense an "incomplete" house. In a type of *nagaya* where a single structure is divided by partitions perpendicular to the direction of the ridge into several units (*munewari-nagaya*), a single unit was apparently only about five square meters in area and the alley too was a meter wide at the most. Even with *nagaya* where each unit is structurally independent, a unit that is over 30 square meters in area is rare (Figs. 4-28, 29). Among today's *nagaya*, the examples in Kyoto shown in Figure 4-30 are approximately 15 to 25 square meters. *Nagaya* in Tsukishima, Tokyo, are also about 25 square meters; however, being two-storied, their total floor area exceeds 30 square meters. In any case, there is not sufficient leeway in a *nagaya* to have something like a *mise* space facing the alley;[27] the entrance is simply a plank on the same level as the *tatami*. There is insufficient family privacy; the family circle is exposed to view from the alley. The units also do not have separate toilets or kitchens; such facilities are often shared.

What is like life for people in *nagaya*? The survey of Shitayama-chō in Tokyo in the 1950s by the British sociologist Ronald P. Dore is helpful. Dore gave the name Shitayama-chō to an actual neighborhood of mixed character that was *shitamachi*-like in some ways and *yamanote*-like in others.

Most houses in Shitayama-chō "present themselves to the [alleys] at odd angles and in higgledy-piggledy order… In the absence of gardens, washing is hung on rows of bamboo poles on special platforms which jut out at first floor level indiscriminately at the front or the back of the house. On fine days the bedding—thick eiderdowns, quilts and nightshirts—is hung from upper windows to air; cooking pots and pieces of furniture are pushed out on to the narrow upper floor verandahs from rooms where living space is insufficient."[28] The surface layers in Shitayama-chō recorded here are endodermal in character and indisputably those we have defined to be the *ura-nagaya* type.

Life in such an environment, particularly "informal neighbour relations," are described as follows. "Few houses have gardens; the disposal of rubbish, hanging out the washing, sitting out in the sun, is as likely as not to bring one into immediate contact with neighbours. The closeness of the houses and the acoustic properties of their wooden walls make it impossible to keep one's prayers or one's parties, one's sorrows or one's quarrels a secret from neighbors… The machinist who went on to expatiate on the good nature of his neighbors remarked that there was no sort of trouble you couldn't discuss with them… Some housewives said that their near neighbors were 'much closer than relatives' and some interviewees commented on the evidence that this was so—sometimes neighbors came into the house and sat themselves

CHAPTER 4   THE EXTERNAL LAYERS OF STREETS   125

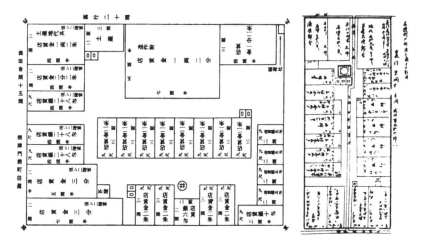

Fig. 4-28 *Left: Ura-nagaya* in Nezu-monzenmachi, Bunsei era (1818–30). From Makoto Takeuchi, *Kansei-Kaseiki Edo ni okeru shokaisō no dōkō.*
Fig. 4-29 *Right: Ura-nagaya* in Kobikichō at the end of the Edo period. From *Nihon Kenchikushizushu.*

1. 6 tatami room
2. 4.5 tatami room
3. Doma
4. Shelves
5. Communal sink

● Potted plant
▭ Washing machine
⊗ Bucket
⊞ Dust pan, trash bin
■ Pickle jar, gasoline can, miscellaneous
— Galvanized iron, plywood, old shoji, wood material
→ Bicycle
Y Faucet
-------- Wash line
▨ Storage
▦ ⊙ ▣ ■ Manhole
▦ Stone Paving

Fig. 4-30 Contemporary *ura-nagaya* in Kyoto. From Shimamura, Suzuka, *Kyō no machiya.*

down during the interview without a knock or an 'Am I disturbing you?'"[29] "Neighbours on such terms help each other in various ways. They act as guardians when someone has to leave the house; for people in Shitayama-chō, though they have an implicit trust in their neighbours, harbor a great distrust of the world outside… They borrow and lend freely. When an unexpected visitor comes and you happen not to have a kettle boiling, it is always worth looking in next-door to see if they have one on the hob. When the gas-man comes too near the end of the month, someone else whose payday was more recent will probably be able to stand in. A household which installs a telephone cannot expect to have exclusive use of it. For every man in Tokyo who has his own telephone number written in the bottom left-hand corner of his name-card, there are three or four who have a neighbour's number with the bracketed qualification 'call-out.' In serious illness help is generous and unstinted. 'People will worry on your behalf as if it was they themselves in trouble,' said the machinist."[30]

This extended excerpt provides a vivid portrayal of people's awareness of the neighborhood in Shitayama-chō. There is little sense of privacy among neighbors; unlike *machiya* or *buke jūtaku*, the family domain is open to the public space. A neighbor is "practically a part of one's family," but the circle of intimacy is not endless; it is limited to the neighborhood. ("People… harbour a great distrust of the world outside.") Such a sensibility is very much a part of traditional comic storytelling (*rakugo*) centered on the lives of the characters Kumagorō and Yatsugorō and is very different from the wariness of neighbors shown by the Nishijin shopkeeper who "looks through the louvers to make sure no one is around before stepping out."

These contrasting attitudes toward neighbors is at times attributed to a difference in temperament between Edo and Kansai (the region including Kyoto and Osaka). However, according to a survey carried out in Kazahayachō, Shimogyō-ku, Kyoto, by the Ueda Research Group of Kyoto University, people living in an alley were on an intimate footing with only others in that alley, and 80 percent of people were more closely acquainted with others in their own alley than with others in the neighborhood generally.[31] Regional differences may impact people's awareness of neighborhood, but differences of class and consequently differences of living environment are apt to have a major effect as well.

As has already been explained, many alleys are culs-de-sac. Those that are not, bend, making it impossible to get a full view of them from the street. There is little through traffic, even though alleys are technically public spaces. Since an alley is used only to access the houses facing it, a stranger there is easily recognized. This is made all the easier by the openness of the houses, which is a result of their small size and the narrowness of the alley. An outsider is met with caution and, if suspicious, is asked what his or her business there is. Entering an alley is like entering someone else's property.

That is, though *ura-nagaya* are individually open to the alley, they are collectively closed in structure to the street. Let us also recall the sense of family solidarity found among occupants of *ura-nagaya*. The alley in the case of *ura-nagaya* can be said to function in the same way as the *mise* spaces of *machiya* and the gate-forecourt spaces of the *buke jūtaku*. To put it another way, the *nagaya* units facing an alley collectively correspond to a single unit of *machiya* or *buke jūtaku*.

## *YASHIKI* AND *NIWA*

*Yashiki* means a tract of land on which a house stands; in the Edo period, the term *machi-yashiki* (town estate) was used as well as *buke-yashiki* (warrior estate). A *machi-yashiki* was occupied in certain cases by a single *machiya*, but often *nagaya* were also developed behind the *machiya*. Nagaya "were created as the dwellings of retainers of *daimyō* beginning in the Kamakura and Muromachi periods when samurai from every province were sent to Kyoto to serve as guards… As castle towns such as Edo, Kyoto and Osaka gradually began to develop in the Edo period, powerful townsmen (*chōnin*) built *nagaya* on their own estates and allowed people who could not live on their own to occupy those dwellings."[32] Eventually, *nagaya* came to be regarded not only as dwellings for one's retainers but as a means of accumulating and making money; *nagaya* generally came to mean rental housing. Since *ura-nagaya* were developed by a landowner on his own estate, *nagaya* facing an alley did not as a rule exceed the size of a *machi-yashiki*. *Machiya* facing a street, on the other hand, were owned by the occupants. They might be occupied by the landowner himself; in many cases the landowner leased just the land, and lessees built houses for themselves. The occupants of *machiya* facing a street were citizens and had civic responsibilities such as compulsory service, but the occupants of *ura-nagaya* were not given the rights of citizens. Thus two types of structures—the relatively large *machiya* with *mise* spaces built along streets and small *nagaya* built along lanes extending into the inner reaches of city blocks—were developed for very different social classes and with contrasting forms in the Edo-period city.[33]

If we look at the relationship between each housing form and its corresponding tract of land (*yashiki*), a single *buke jūtaku* or a single *machiya* could occupy by itself a tract of land; in the case of *ura-nagaya*, a tract of land was occupied by a cluster of such units organized around an alley.

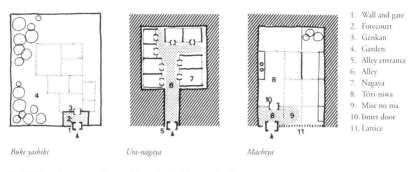

*Buke yashiki*    *Ura-nagaya*    *Machiya*

1. Wall and gate
2. Forecourt
3. Genkan
4. Garden
5. Alley entrance
6. Alley
7. Nagaya
8. Tōri-niwa
9. Mise no ma
10. Inner door
11. Lattice

**Fig. 4-31** Spatial structure of external layers in feudal urban dwellings.

That is, every form of housing at the *yashiki* level possessed a spatial device for controlling the relationship between the inside and the outside of the *yashiki*—namely, gate-forecourt-*genkan*, *mise* spaces and alley—and was an autonomous domain despite being located in the city (Fig. 4-31).

In this way, the *yashiki* was an autonomous unit of landownership and life. It was also a unit of political control and, if we consider the belief in *yashiki-gami*, a religious unit as well. Even today, a small shrine (*hokora*) is found in a corner of most farmhouses and large estates, and the presence of Inari shrines on the roofs of many tall buildings shows that this belief in *yashiki-gami* still survives. The belief in *yashiki-gami* has the character of both ancestor worship and worship of the *kami* of place or locality (*tochigami*) and takes diverse forms, but in all cases a *yashiki-gami* serves to protect a certain tract of land."[34] A *yashiki* was land that ought to be protected; it was the area over which the *kami* could exert its spiritual influence and was considered a closed domain.

Thus the *yashiki* was a compositional unit of the city, a domain that was autonomous in character as far as the spatial mechanism of the house occupying it, ownership and religion were concerned. Furthermore, all *yashiki*, be they the *yashiki* of *buke jūtaku* or the *yashiki* of *machiya*, tended to be used spatially in the same way—that is, buildings on all *yashiki* tended to be arranged in the same way.

The main residence in Edo of the Hikone domain (Fig. 4-32) was arranged with the residence of the *daimyō* and his wife in the center, surrounded by *nagaya* of different sizes for retainers and servants on the perimeter of the site. Between the central residence and the peripheral *nagaya* were outdoor spaces. Such outdoor spaces within a *yashiki* were called *niwa* in ancient times, and gardens, which are today referred to as *niwa*, were then referred to as *shima*, written as 山齋 or 山池. *Niwa* was then written as 屋前, 屋外, or 屋戸 and signified spaces surrounding a residence.[35] As the history of Japanese architecture makes clear, buildings placed in *niwa* developed free, irregular forms. That became a major characteristic of Japanese architecture and widespread even in ordinary houses. *Niwa* therefore do not have clear geometrical outlines; they are residual spaces, left over by buildings.

*Niwa* take the form of *roji* in *machi-yashiki*, and they too have a residual character. The external layers of *ura-nagaya* display the general characteristics

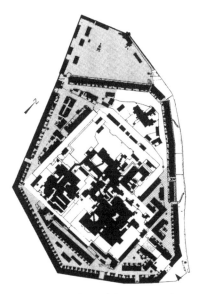

**Fig. 4-32** *Niwa* in a *buke-yashiki*; the main residence in Edo of the Hikone domain. (Based on a drawing by Kōji Nishikawa.) In a large residence such as this, the grounds were surrounded not by a single wall but by *nagaya*—that is, a deep zone (shown as a meshed area), where retainers and servants lived. If we examine just the *nagaya* in the meshed area, they are arranged according to the same principles as *nagaya* in *machi-yashiki*.

**Fig. 4-33** Contemporary *nagaya*. After World War II, the Housing Corporation constructed housing projects and supplied rental housing in cities all over Japan. It also experimented with new housing forms, such as the terrace houses in the Asagaya Housing Project developed in a western suburb of Tokyo. Perhaps because the housing is a form of *nagaya*, there are also *niwa*-like characteristics in the way exterior spaces are used. The boundaries of gardens of units are irregularly arranged.

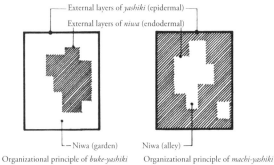

Fig. 4-34 Spatial concepts of *buke yashiki* and *machi yashiki*. In schematic form, one is seen to be the negative of the other.

of building arrangements facing *niwa* in a *yashiki* and are indeed endodermal (Fig. 4-33). That is, buildings facing *niwa*, whether in a *machi-yashiki* or *buke-yashiki*, possess endodermal external layers. Similarly, the external layers of *yashiki* on the public space side are in all cases epidermal, demonstrating that *yashiki* are the formal units of the city. That is because the epidermis is the membrane enveloping an autonomous, complete organism. *Ura-nagaya* are in contact with public space only at the entrance to the lane and have practically no "epidermis." That is a clear indication that this housing form was in a position that was socially and formally subordinate.

All building forms facing *niwa* are endodermal in character, but there is a difference in the treatment of plants, that is, nature. *Buke jūtaku* captures nature within an enclosure in the city, and buildings are placed inside that nature. *Niwa* are gardens surrounding buildings where trees are planted and nature is simulated. On the other hand, *niwa* in the case of *ura-nagaya* are passageways—artificial spaces surrounded by buildings. Extremely artificial forms of nature—potted plants—are placed in an artificial environment and admired (Fig. 4-34). The *doma* of *machiya* may be called a *tōri-niwa* (literally "pass through *niwa*") because it too is a passageway and an artificial space. The idea in a *buke-jūtaku* was to introduce a man-made order into nature, and in a *machiya* it was to introduce nature into a man-made order. The two contrasting treatments of nature within *yashiki* seem to reflect the two systems or traditions of Japanese dwelling forms discussed by Hirotarō Ōta:[36] in one the floor of the dwelling is separated from the ground (nature), and in the other the floor of the dwelling is hardened earth.

## TYPES OF EXTERNAL LAYERS

The external layers arranged between the "inside" of a house and the street do not reflect the internal functions of the house; instead, their form is created by spatial mechanisms for reconciling two demands, seclusion and contact with the world outside the house on various levels, including the maintenance of social relationships and the interior climate of the house.

Most of the residential areas we have studied were not designed by someone in a uniform way but were created in different periods for people of different means and tastes. The fact that they all converged nevertheless on four types indicates that as with the layouts of houses, people conceived the external layers of houses as "types." A type comes about when a number of relationships between formal elements are considered an inseparable set and incorporated into a semantic structure based on historical origin. The three forms of urban dwelling that developed in feudal times correspond to the class structure in cities in feudal society, and the forms of their external layers are indeed the "faces" of dwellings and symbolized the social position of their occupants. Needless to say, the external layers of *buke jūtaku* were of the highest status.

*Buke jūtaku* were more than just the residences of samurai, the ruling class in feudal times. Samurai had exclusive rights to that dwelling form; in principle, no other class was permitted to use it. The incorporation of gates and *genkan*, that is, elements of the external layers of *buke jūtaku*, was forbidden in the dwellings of townspeople and peasants; those forms were objects of desire of commoners. In public, people welcomed Westernization in the Meiji period, but the ethics of the samurai class continued to be the model in private life. The new ruling class of bureaucrats, military officers and entrepreneurs adopted the *buke jūtaku* as their form of dwelling, and the public too aspired to it. That situation continued and indeed spread after World War II; the

Fig. 4-35 Meiji-period house. Sketched by Edward S. Morse, *Japanese Homes and Their Surroundings*.

general public cannot seem to imagine other forms of dwelling. Even a house on a small site for an ordinary office worker is surrounded by *niwa*; a wall is built on the boundary of the site; the convention of having a gate-forecourt-formal entrance is observed, though the elements are reduced in size to fit a narrow site (Figs. 4-16, 35). As a result, the main building is revealed to the street. External layers of a type entirely different from those of the *oyashiki* type have developed and shape the mainstream of suburban houses today. They are not as ceremonial as external layers of the *oyashiki* type, nor do they promote as intimate a relationship with neighbors as external layers of the *ura-nagaya* type do. External layers of this type instead project the image of a middle-class lifestyle, one in which moderate social intercourse is enjoyed across hedges.

As this tendency for houses to become smaller progresses, the external layers of the suburban house type approach the external layers of the *ura-nagaya* type. Since the external layers of the main buildings were originally an endodermis wrapped and concealed in an epidermis—that is, the external layers of the main buildings were arranged according to the *niwa* principle—and now are being exposed, they possess something of the character of the external layers of *ura-nagaya*. The miniaturization of suburban houses (as in so-called mini-developments), however, strips the main buildings of the last vestiges of *niwa* (Fig. 4-36). Houses of this sort, however, are not situated on quiet alleys, access to which is limited. They present a pitiful sight—like an endodermis showing from an open wound.

In addition to the formation of the suburban house type, various other instances exist in which external layers have been interpreted as a type and a social predilection for the *oyashiki* type is revealed. Let us compare the garden walls surrounding a condominium in Komaba[37] and a detached house in

**Fig. 4-36** *Left:* An example of a type of high-density development of small, ready-built houses known as *mini-kaihatsu* (mini-development); Nishi Ogikubo, Suginami-ku, Tokyo.
**Fig. 4-39** *Right:* The boundary as symbol. In this house in Nishikata, Bunkyō-ku, Tokyo, a fence is built to mark the boundary of the site, even though the windows of the reinforced concrete house are already equipped with protective grills.

Nishi-Ogikubo[38] (Figs. 4-37, 38). A similar form of property wall is adopted, even though these walls surround structures of an entirely different nature. The fence in Nishikata is necessary to the house's image as "a house with a property wall," even though it serves no practical function (Fig. 4-39). External layers of the *oyashiki* type are used not only for houses but for every type of public institution. The Hongō Campus of the University of Tokyo has external layers of the *oyashiki* type at its boundary with the city, even though its buildings, modeled on those of British universities, are in the neo-Gothic style (Figs. 4-40, 41).

The results of our study of forms of dwellings in the feudal city can be added to the diagram of types of external layers already indicated (Fig. 4-42). First, the vertical axis can be reinterpreted as showing the opposition between the external layers of the *yashiki* and the external layers of the *niwa*. The axis indicating the opposition between cases where the interface with neighboring houses is the second layer and cases where the interface is the first layer overlaps the opposition in ways nature is introduced. That is, with a type

Fig. 4-37 External layers of a condominium; Tomigaya, Shibuya-ku, Tokyo.
Fig. 4-38 External layers of private houses; Nishi Ogikubo, Suginami-ku, Tokyo.

Fig. 4-40 External layers of the Hongō campus of the University of Tokyo.
Fig. 4-41 External layers of Oxford University.

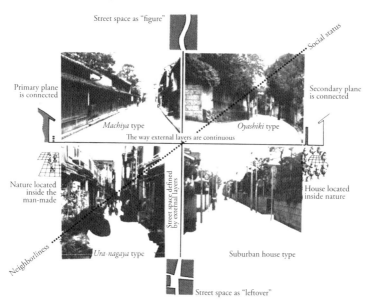

Fig. 4-42 The semantic structure of external layers.

where the first layer is continuous, the main building is surrounded by *niwa* and located amid nature. On the other hand, with a type where the second layer is continuous, the *niwa* is enclosed in a building, and plants are placed in a man-made environment.

The four types of external layers are models that individually can be distilled logically, and types of external layers of similar character may be found in other cultures. However, the coexistence of these four types in a city, and the existence of a significant third, diagonal axis extending from the top right to the bottom left in this diagram are unique to Japan since the feudal period. This diagonal axis not only expresses the volume of communication between the inside of a house and the street, that is, the qualitative difference in neighborhood relationships determined by external layers, but by relating to the size of sites, corresponds to class hierarchy. This is unique to Japan because houses in which the external layers of main buildings are arranged along the street do not necessarily demand small sites. For example, as with Italian palazzos, there is a tradition in the West of creating a garden in the back and constructing the building along the street even on extensive sites. Therefore, this diagonal axis indicates a social semantic structure between types of external layers in Japan, that is, a predilection for external layers of the *oyashiki* type. As we have already seen in the example of a university campus, that semantic structure extends to the external layers of public buildings as well as houses. This tendency to give even public, monumental buildings that are urban landmarks external layers of the *oyashiki* type makes Japanese urban spaces different from those in the West. The external layers of the *oyashiki* type are based on the principle that the more formal or ceremonial its character, the further the main building ought to be constructed from the first layer. Important buildings do not participate in urban spaces. One sees only high walls and tree-lined streets. One could say that the most monumental structures in Japanese urban spaces are gates. Buildings in Western cities stand along the street, and their degree of importance is directly expressed by their size or height. In Japan, however, the gate is quite symbolic, and the style of the gate expresses the degree of importance of a building.

## 2. THIN LAYERS AND GAPS

The above has been a description of the external layers of residential areas in Tokyo and Japan generally. However, the formal characteristics of those external layers need to be more fully described if we are to account for the distinctive nature of the Japanese urban environment. For example, *machiya* can be said to form a continuous wall surface that enables the street space to be read as figure, but in truth such a characterization would more accurately apply to the wall surfaces of an Italian city. Using the Yoshijima House (Fig. 4-44) as an example, let us examine in more detail the external layers of *machiya*. First, there is a ditch approximately 40 centimeters wide along the boundary of the site, and the main roof extends to a point almost directly above the ditch. The wall of the building proper is set back approximately one meter from that point. Louvers are attached to the outside of that wall. In addition, the area between the wall and the ditch that is covered by the roof overhang is enclosed by a short fence (*komayose*). Four parallel layers thus compose the outer limits of this house. Each layer is porous, and it is not clear precisely where the inside of the house begins. However, in an Italian city built in the medieval period, one layer clearly marks the boundary between the inside and the outside of a house. That layer is so dense and solid it lets nothing through except by way of doors and windows.

Rowhouses in the eastern United States or England which can be compared to *machiya* in Japan, have similarly to *machiya*, external layers (Fig. 4-45) that are spatial zones between the buildings and the street. However, their external layers produce a quite different impression. In short,

Fig. 4-43 *Top:* External layers of San Gimignano, Italy.
Fig. 4-44 *Bottom:* External layers of a *machiya*; the Yoshijima House (photographer: Akiyoshi Imakura).

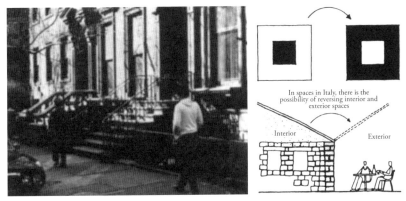

**Fig. 4-45** *Left:* External layers of rowhouses in New York.
**Fig. 4-46** *Right:* The reversal of inside and outside in Italian architecture. "Italian streets and piazzas are beautifully paved in their entirety and not much different from interior floors. Masonry walls separating building interiors from the street are also treated much the same, inside or outside. The only difference is that the interiors are roofed. Thus the spatial organization in Italy, based on a perception that the outside has many qualities in common with the inside, permits the outside to be seen as figure." From Yoshinobu Ashihara, *Machinami no bigaku*.

the street space created by the external layers of a rowhouse is substantial, almost tangible, in character, whereas that created by the external layers of the *machiya* type is tenuous. The architect Yoshinobu Ashihara has explained the nearly tangible character of street spaces in the West using the concept of figure and ground from Gestalt psychology (Fig. 4-46).[37] "Italian streets and piazzas can be read as clearly outlined 'figures'... That is possible because buildings form a continuous plane along the street. An isolated, monumental building naturally assumes a more prominent role, in which case the street then becomes the 'ground' to that 'figure.'"[38]

The external layers of other building types beside the *machiya* also produce ambiguous boundaries and tenuous street spaces. The preceding section focused mainly on the distinctions between different types of external layers in residential areas of Tokyo. In this section, we will examine formal characteristics that are common to all types of external layers and that make Japanese cities distinctive.

The different impressions produced by the external layers of Japanese and Western cities may be explained by differences in the treatment of layers. Since layers (such as walls and partitions) are the boundaries between spaces, differences in the treatment of layers can be said to result from different modes of spatial perception. For our analysis we will therefore use evidence drawn from not simply the external layers of buildings but layers found in everything from architectural interiors to cities.

Fig. 4-47 *Top: Shinkabe* construction, Myōshinji, Kyoto.
Fig. 4-48 *Center:* Kenninji-style bamboo wall; Shōan, Suginami-ku, Tokyo.
Fig. 4-49 *Bottom: Noren*, Kyoto.

## THIN LAYERS

Layers in Japanese cities and buildings, particularly traditional cities and buildings, seem thin. First of all, the layers are literally thin in cross-section. *Machiya* are basically of *shinkabe* construction (Fig. 4-47). Walls are built between columns; the columns are thus exposed on the exterior. That means nothing in these buildings exceeds the columns in thickness. There is certainly nothing comparable to the walls tens of centimeters in thickness found in Western masonry buildings. Moreover, many features of these buildings such as battened ceilings (*saobuchi tenjō*) and battened walls (*yamatobei*) are designed expressly to show the thinness of members.

Thin means flimsy. The word *hei* (meaning a wall on the perimeter of a lot) was introduced into Japan with the arrival of Buddhism; until then, there had been only the word *kaki* (meaning a fence or hedge).[39] One can peer through a *kaki*.[40] A *kaki* is thin and fragile (Fig. 4-48). Even the *sujibei*, the wall having the sturdiest construction and indicating the highest social status among traditional walls, was at most only four meters high. As a perimeter wall for an imperial residence it seems to provide inadequate security. Boundaries within a residence such as *shōji* (sliding screens of translucent paper on a wooden lattice), *fusuma* (sliding doors or partitions of paper on a wooden frame), *sudare* (rattan blinds) and *noren* (short curtains) are all thin; their treatment requires great care.

Secondly, thin means light in weight. Hedges, board fences (*itabei*), *shōji*, *fusuma* and *noren* (Fig. 4-49) all seem light and are in fact light in weight. Western doors are heavy and apt to creak when opened. *Fusuma* and *shōji* of good quality slide open soundlessly and effortlessly. The Japanese prefer light things over heavy things. These objects, being light, are easily demounted. The Japanese are accustomed to replacing one form of light boundary with another

Fig. 4-50 *Left:* Louvered screen made of logs; Gokasho, Kanzaki-gun, Shiga prefecture.
Fig. 4-51 *Right:* Shadow of louvers on *shōji*; Yoshijima House.

to suit the season or the mood. There are many examples of such boundaries, such as the *sudo* (reed screens) used in summer, *manmaku* (curtains) hung at flower-viewing parties and athletic meets and *tsuitate* (single-leaf screens) used to create a room within a room.

Thirdly, thin means capable of letting through light. There are numerous examples of lattices (Fig. 4-50) used to form boundaries in traditional Japanese architecture such as the various lattices (*kōshi*) in *machiya* and louvers (*renji*) and "lath windows" (*shitajimado*) in *sōan* (thatched-hut) tea house. Slats can be spaced closely together to control light and still let a person on the darker side of the louvers know what is happening on the other side. Scenes glimpsed through *sudare* and shadows cast on *shōji* (Fig. 4-51) appeal to the Japanese aesthetic. One might even go so far as to say layers in Japan are used not so much to divide or isolate but to provide people with the pleasure of looking through them.

Layers that are thin, lightweight and porous are naturally poor in sound insulation. Thin layers do not adequately protect or block. Thin layers therefore can function as boundaries only if certain social conventions are observed. That is, the layers are largely symbolic in character. The external layers of Japanese cities and buildings are not just physically thin. The elements from which they are composed are formally related to each other in such a way as to suggest thinness. That is, the layers are designed to emphasize the fact that they are thin. In Japan, layers are not created by massive walls but composed of, and ornamented by, linear elements. Everything from lattices to walls of *shinkabe* construction is composed of straight lines meeting at right angles. Diagonals and curves are generally eschewed. Moreover, as can be seen in lattices of the feudal period, care is taken to discriminate between elements that intersect. That is, the elements are placed on slightly different planes or given slightly different widths; squares are avoided. As a result, each linear element possesses directionality; straight lines seem to extend, unimpeded, over the plane. Louvers in which slats of the same dimension meet flush, that

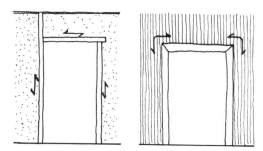

Fig. 4-52 Comparison of frames with miter joints and horned joints.

 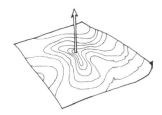

**Fig. 4-53** *Left:* Door and windows of a palazzo; Palazzo Cicciaporci by Giulio Romano. From Peter Murray, *Renaissance Architecture.*
**Fig. 4-54** *Right:* Contour lines on a map. Contour lines are not simply a convention; they rely on the perceptual qualities of concentric lines to suggest height.

is, in the same plane, were not much used in residences of the feudal period, except in conservative institutions such as temples and the imperial palace. In such louvers, directionality is erased; the composition is uniform and static. Swastika patterns of bent, intersecting lines, such as those found in Chinese louvers, are also rare.

A desire to emphasize the directionality of each intersecting element and to avoid lines that turn corners is also evident in the treatment of door and window frames; there is a tendency to let one element run past the other instead of having them meet in a miter joint. A frame with miter joints encloses the door or window opening and gives it unity and completeness. On the other hand, in a frame with "horned joints" (*tsunogara*) where one element is extended beyond the other, lines do not turn corners; each line retains its own directionality. A concentric pattern is often used in frames in the West (Fig. 4-53). Such a pattern—which is also used in the design of the door panel itself—underscores the formal qualities of a frame with miter joints. Such patterns are similar to contour lines used to indicate mountains and valleys on maps. Just as contour lines are an expression of the third dimension on the flat plane of a map (Fig. 4-54), the concentric frame pattern used in the West is a way of suggesting a direction perpendicular to the plane that has been ornamented with the pattern; that is, the pattern is a way of suggesting thickness.

In Japan a frame with horned joints does not turn corners; vertical and horizontal elements try to go their own ways (Fig. 4-52). The directionality expressed by intersecting lines in Japan is contained within the plane. The lines suggest infinitely extending forms, which produce the effect of a surface that is under tension. In contrast to Western architecture, the actual thickness of the layer is erased.

## LAYERS THAT SCREEN OFF SPACE

The thin layers of Japan are without exception flat; they are never bent or curved. In the Yoshijima House, which we examined earlier, the wall built up to the street on the right-hand side seems at first glance to turn a corner. However, the south wall and the east (streetside) wall differ in height and finish; the east side is also different in that it is roofed. Thus, the effect is of two separate layers instead of one layer turning a corner. A similar relationship can be seen in the clay wall (*tsuijibei*) of the temple Myōshinji in Kyoto. Since the wall encloses the compound of a *tatchū* (quarters established to tend the grave of a renowned priest), that is, since it encloses a single, well-defined area, one would expect the wall to be continuous, like a string strung along the boundary line. However, the treatment of the corner suggests two separate walls coming together; one wall is terminated, and its end is revealed (Fig. 4-55). The relationship between two layers is thus similar to the relationship between two intersecting linear elements within a layer. Spaces defined by layers that do not bend might be better described as screened off rather than enclosed.

The tendency not to bend layers is also found in the design of interiors. In interiors with vaulted or domed ceilings in many parts of the world, the ceiling and the wall are often treated as one continuous curved plane. In Japan, however, they are most often treated as separate planes. In *sukiya* architecture, the tendency is found even in the relationship between walls. Horizontal members are staggered so that they do not meet at corners. A room therefore is a collection of independent planes.

This avoidance of any formal treatment that may suggest the enveloping of space is also evident on an urban scale. The entrance to a *machiya* is usually on a side of the building that is parallel to the roof ridge. Even in a *machiya*

Fig. 4-55 Corner of a *tsuijibei*, Myōshinji.

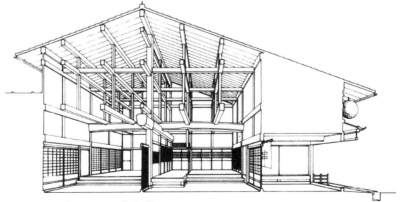

Fig. 4-56 *Top: Machiya* on a corner lot in Kyoto.
Fig. 4-57 *Center:* Longitudinal cross-section of St. Philibert. From Kenneth John Conant, *Carolingian and Romanesque Architecture 800–1200.*
Fig. 4-58 *Bottom:* Longitudinal cross-section of the Yoshijima House.

standing on a corner lot, the eaves rarely turn the corner (Fig. 4-56).[41] In Western cities, corner situations were acknowledged architecturally, for example, by locating the main entrance or a tower at the corner or rounding off the corner. In Western cities, building facades constitute the enveloping outer skin of city blocks.

The man-made physical environment, both architectural and urban, is intended to give order to space. Delimiting territory is one way of giving order to space, and establishing a territorial hierarchy is another. Let us compare the Yoshijima House and the Romanesque church of St. Philibert in Tournus. The place furthest from the entrance is occupied in the former by the family room and in the latter by the altar. The two buildings are of course different in character, but in both cases space is rhythmically articulated and a hierarchy of ascending order is established in the direction of that furthest place. However, space is articulated in markedly different ways.[42] In St. Philibert, the transverse arches and the tunnel vaults arranged at right angles to the main axis of the church endow each bay with centrality and unity. The spatial experience here is like passing through several rooms (Fig. 4-57). In the Yoshijima House, on the other hand, transverse beams and three sets of horizontal members and struts located above the beams work together to create a series of thin layers that are perpendicular to the movement toward the inner depths of the building. However, there is none of the unity or centrality of space found in St. Philibert. The spatial experience is like passing under a series of curtains (*noren*) (Fig. 4-58).

## OVERLAPPING THIN LAYERS

Overlapping thin layers not only establish a hierarchy but produces an

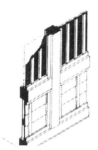  

Fig. 4-59 *Left:* Detail of window on the front elevation of the Yoshijima House.
Fig. 4-60 *Center:* Typical arrangement of the front door in an *oyashiki*-type house; Minami Ogikubo, Suginami-ku, Tokyo. The gate is displaced from the axis of the front door.
Fig. 4-61 *Right:* Overlapping layers; Shimogamo Shrine, Kyoto.

impression of spatial depth. This is a method of organization widely used in external layers and architecture in Japan. I have already described at the outset the way the external layers of *machiya* are organized, and the same method is used in details as well. In the Yoshijima House the windows to the left of the entrance are composed of three layers: louvers, blinds and *shōji* (Fig. 4-59). They are all thin layers but different in character. Louvers let in light and air and are solid. Blinds are flimsy but prevent people from looking in from the outside during the day. *Shōji* screens disperse light and are easily opened or closed. The idea here was to meet the complex demands made on windows on the streetside of the residence by combining layers of different character.

Many widely used techniques in architecture and garden design involve the overlapping of thin layers. If one layer is arranged directly in front of the layer behind it, the overlapping of the layers is not apparent to an observer. In feudal *shoin*-style residences, therefore, buildings were typically arranged in a staggered fashion. When the shoin of Katsura Detached Palace in Kyoto is viewed from the garden, the layers represented by the walls and the *shoji* of the Old Shoin, Middle Shoin and the New Shoin and the layers represented by the verandas and posts are seen to be arranged in parallel, creating a sense of depth and rhythm.

The widely observed practice of locating the gate off the axis of the front door in large residences and suburban houses can be regarded as a variation on "staggering" (Fig. 4-60). In these cases there are also two layers, the fence and the wall of the main building; the displacement of the gate from the axis of the front door visually underscores the overlapping of the two parallel layers. The displacement also forces a change in movement. Movement parallel to the layers is added to movement at right angle to the layers; the parallel arrangement is experienced as well as seen.

"Screening" (*saegiri*), one of the ways used to interrupt the view and thus provide only intermittent glimpses (*miegakure*) of an object in the landscape, and "borrowed scenery" (*shakkei*), the incorporation of a distant view into the landscape in the foreground, can be interpreted as ways of creating an impression of spatial depth by the overlapping of thin layers. The Japanese have traditionally found aesthetic delight in such spatial relationships (Fig. 4-61).

## OBJECTS AND SPACES AS VOLUMES

Surfaces in the West suggest, by their physical presence and design, that there is something behind them. They are thick. They are plastic membranes enveloping space. A difference in spatial perception may be behind the

Fig. 4-62 *Left:* Stained glass; Sainte Chapelle, Paris.
Fig. 4-63 *Right:* Chapels of a pilgrimage church. From C. Norberg-Schultz, *Late Baroque and Rococo Architecture.*

difference in the treatment of layers in Japan and the West. In the West, objects and spaces have been regarded as cohesive volumes. Layers have been seen as membranes enveloping those volumes. The cultures of Egypt and Greece emphasized the cohesive quality of objects but did not see spaces as volumes. Space began to be expressed as volume in the West with the use of stained glass in Gothic cathedrals. With stained glass, "light was made manifest as spatial volume. The result was the flooding of the interior with light. Particles of light filled space."[43] Stained glass is itself a thin layer materially, but its pattern, for example, the pattern of a rose window, possesses centrality. Moreover, the stained glass windows in an ambulatory with radiating chapels encircle space. Stained glass panels, in relationship to the building as a whole, serve as a plastic membrane enveloping a volume of interior space (Fig. 4-62). The Baroque expressed through architectural massing the competing pressures within the city—the pressures of the interior spaces of houses and the exterior spaces of streets and plazas. Baroque buildings are either pushed out by internal pressure, until walls finally expand between and beyond towers (Fig. 4-63) or, as in the piazza of S. Ignazio in Rome or the Royal Crescent in Bath they are pushed in and bent by the pressure of the exterior spaces of cities.

The idea of spaces and objects as volumes was adopted on an urban scale in medieval walled cities. Those cities were based on a dualism that pitted cosmos against chaos, sacred against profane;[44] the city had to be covered by a network of order created by humankind. Order had to be imposed on a medieval Italian city; every corner of the city had to be filled with coherent, complete volumes of either space or matter. This idea is most eloquently expressed by the well-known map of Rome by Nolli. As has been pointed out,[45] both buildings and exterior spaces can be perceived as figure to the other's ground. As I have already explained, plastic membrane-like layers that suggest the

Fig. 4-64 *Left:* Plants growing in a gap; Shiba, Minato-ku, Tokyo.
Fig. 4-65 *Right:* Urban gap; *Shōheizaka, Seidō, Kandagawa* by Ando Hiroshige, woodblock print.

presence of something behind them make palpable the idea of objects and spaces as volumes.

## GAPS AND NATURE

The thin layers of Japan, on the other hand, deny their own material thicknesses and do not act on spaces in front of or behind them. Forces and directionality expressed on a layer may endow that layer with tension but never transcend the layer. The layer is isolated. Then what exists between one layer and another? What is the relationship between layers?

There is an implicit belief that natural order rather than human order prevails between one layer and another. That is, the order imposed by humankind is concentrated in the layers themselves. No effort is made to impose human order on the intervals between layers. Such intervals are left in their natural or quasi-natural state.[46] Walking around a Japanese city, one sees grass and trees springing up everywhere between man-made objects (Fig. 4-64). The Japanese prefer not to cover the earth entirely with man-made objects. Buildings do not rest on podia but are raised up in the air by posts and struts so that the area underneath the floor is left in its natural state. Roads are not heavily paved. The paths running between the *tatchū* of Myōshinji are paved only in the middle; between the paving and the walls on either side of a path is a zone of bare earth and plants. I have stated that Japanese boundaries are intended only to screen off. The walls of the *tatchū* at Myōshinji merely screen off land that is natural; even after the land has been screened off, nature remains on either side of the partition.

In Japanese, intervals between expressions of man-made order are referred to as *sukima* (literally, "empty spaces"). If to fill in completely is Western, then to leave such gaps is quintessentially Japanese. Gaps can be found on an urban scale. Edo at its peak was the largest city in the world in population. However, as can be seen in woodblock prints, there were places, even in central Edo, that were left wooded (Fig. 4-65). There were gaps in cities, and nature revealed itself in such gaps. On the other hand, gaps left between buildings determine the quality of the external layers of Japanese cities. In the feudal period, no freestanding, multi-unit form of housing ever developed other than the *nagaya*. Even the *machiya*, which had similar heights and similar elements such as louvers, were detached houses having no shared walls. As a result, narrow gaps were left between buildings. Looking at the streetscape produced by a series of *machiya* in Takayama or Kyoto, one sees that the lower edges of roofs are not perfectly aligned—they come forward or recede a little from house to house—and gaps are produced everywhere. The overlapping of layers is interrupted time and time again, leaving gaps between layers. Unlike the external layers of medieval cities in Italy, the external layers of Japanese cities do not seal in the air of the street space. The Japanese prefer to establish a relationship between objects that permits, as it were, the slow, steady circulation of air.

As I stated at the outset, this analysis is not limited to external layers, but I believe what has been described explains the characteristics of the external layers of Japanese cities. These characteristics, which can be summarized as "thin layers and gaps," can be seen in the four types of external layers described in the previous section and contribute to the distinctiveness of Japanese cities. Of course, they do not fully explain the characteristics of the external layers of cities and the boundary surfaces of buildings or the spatial concepts of Japan. However, "thin layers and gaps" are undoubtedly an important aspect of the formal and spatial concepts that have prevailed in Japan since the feudal period.

NOTES

1. Topographically, Tokyo at its eastern, bayside end is a low-lying area; from there, it extends westward. In the Edo period, the *chōnin* (artisans and tradesmen) district was located to the east of Edo Castle (today's Imperial Palace), and the estates of the samurai, the ruling class, were located around the castle and on uplands to the west. That is, Edo Castle was at the edge of a plateau and located in a strategic position overlooking the low-lying commercial district. In the Meiji period, the uplands came to be occupied by government officials and military officers. Suburbanization, which began in the 1920s spread to the west, creating residential areas for the new middle class of office workers.
2. *Yashiki* means an estate, a tract of land on which a residence is built. The honorific "o" added to *yashiki* denotes respect and implies acknowledgment of high status. *Machi* can mean "town" or, as in this case, "neighborhood." As will be explained, the upscale *oyashiki* of modern times is modeled on *buke jūtaku*, the residence of samurai, the ruling class of the Edo period.

3. In this essay, "house" (*jūkyo*) refers to not only a building but the entire site as well. The site is the domain of the family, and the way the building is arranged on the site can entirely alter the character of a dwelling and also influence the form of the building.
4. This study was undertaken in 1978. Car ownership had become widespread among Japanese households, and the Capital Region was expanding further. However, there were as yet only nine buildings in Tokyo exceeding 150 meters in height. Tokyo subsequently underwent a major transformation, and some of the characteristics observed here have now become less apparent.
5. Tachikawa is located 50 kilometers west of Tokyo. In 1922, it became the site of the Tachikawa Air Field of the Japanese Army; after the war, it developed into a US military base. Today, it is a core city in the western suburbs of Tokyo.
6. A *shimotaya* is a former shophouse that has closed its business.
7. Alleys that give access to *ura-nagaya* are called by different names depending on the region. In Edo, they are called *nuke-ura*, *fukuro-kōji* and *roji*; in Osaka, *roji* or *rōji*; in Nagoya, *kansho*; in Kyoto, culs-de-sac are called *roji*, and alleys that are open at both ends are called *tsuchiko* or *zushi*. Uzō Nishiyama, *Nihon no sumai* (Dwellings of Japan) (Tokyo: Keisō Shobō, 1976), p. 66.
8. At the time of the study, many *ura-nagaya* were still to be found in lanes, but subsequent rebuilding has given dwellings external layers similar to those of suburban houses.
9. Vernacular houses in Sotodomari, Kōchi prefecture, stand on slopes; they are surrounded by stone walls on the seaward side, and the buildings are of the courtyard-type in plan. There is a gap between the wooden exterior wall of a house, which has few windows, and the stone wall. Its external layers are a deformation of the residence type. See *Toshi Jūtaku* (Tokyo: Kajima Institute Publishing Co., Ltd, 1979) p. 36.

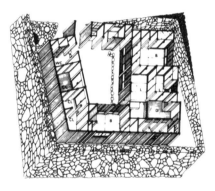

10. Edward Hall, *The Hidden Dimension* (New York: Doubleday, 1966). From observations of animal behavior, Hall developed ideas of "personal distance" "social distance," and "public distance." He declared that there was a cultural dimension to such distances; it should be kept in mind therefore that the distance cited here was the result of a study of Americans.
11. Streetscapes of *machiya* had almost vanished in Tokyo by 1978, when the author wrote the article, but I have chosen to include it as the type is indispensable to a discussion of urban dwellings in Japan.
12. See, for example, Atsushi Ueda's discussions of *machiya* in Atsushi Ueda and Atsuo Tsuchiya, eds., *Machiya—kyōdō kenkyū* (*Machiya*: Collaborative Studies) (Tokyo: Kajima Institute Publishing Co., Ltd, 1975), p. 5.
13. In farming villages the house type is known as *nōka*. The influence its external layers had on urban dwellings is not clear but may account for the presence of hedges and groves (*yashiki-rin*) around suburban houses.
14. To qualify for the title of *daimyo*, a fief-holder's land had to yield at least 10,000 *koku* (approximately 80,000 bushels) of rice annually. One koku was considered enough to feed one person for a year.
15. With the Meiji Restoration, and as the government espoused a policy of Westernization, public buildings and private-sector buildings that were public in character such as banks and department stores began to be constructed in Western styles. The streetscapes of Tokyo took on what might be called a colonial appearance, with a mixture of Japanese and Western styles. The affluent adopted a split lifestyle—with everyday life conducted in a Japanese-style building and guests received in a Western-style dwelling—and this eventually influenced the design of suburban houses of the middle class. By the 1920s, a style of dwelling developed in

which a reception room that was Western in its interior and exterior finish was built beside the entrance to a Japanese-style main building.
16. Naomichi Ishige, *Jūkyo kūkan no jinruigaku* (The Anthropology of Residential Spaces) (Tokyo: Kajima Institute Publishing Co., Ltd, 1972).
17. Ishige, *Jūkyo kūkan no jinruigaku*, p. 246.
18. Hara Research Group, Institute of Industrial Research, University of Tokyo, "Jūkyo shūgōron sono 1— Chichūkai chiiki no ryōikironteki kōsatsu" (Housing Set Theory 1: Investigations into Territoriality in the Mediterranean Region), SD Supplement, no. 4, Kajima Institute Publishing Co., Ltd, 1973.
19. Hara Research Group, *Jūkyo shūgōron*, p. 33.
20. Serge Chermayeff and Christopher Alexander, *Community and Privacy: Toward a New Architecture of Humanism* (New York: Doubleday, 1963).
21. Chermayeff and Alexander, *Community and Privacy*, pp. 216, 218.
22. Wajima is a small city on the Noto Peninsula facing the Sea of Japan in the middle of Honshu, the main island in the Japanese archipelago. Its principal industry is fishing, and it has long been famous for its lacquerware.
23. See, for example, Ueda and Tsuchiya, *Machiya—kyōdō kenkyū*.
24. For a detailed discussion of the Yoshijima House, see Hidetoshi Ohno, "Yoshijimake jūtaku no keitai, kōzō to sono imirontenki kankei" (Formal Structure and Semantic Relationships in the Yoshijima House), *Nihon kenchiku gakkai ronbun hōkokushū*, no. 278, 1979.
25. Atsushi Ueda, *Kyō machiya—komyūnitii kenkyū* (Kyoto Machiya: Community Studies) (Tokyo: Kajima Institute Publishing Co., Ltd, 1976), p. 27.
26. Uzō Nishiyama, *Nihon no sumai*, p. 36.
27. R.P. Dore, *City Life in Japan: A Study of a Tokyo Ward* (Berkeley: University of California Press,1958). This study was carried out in 1951–54.
28. Ibid., p. 16.
29. Ibid., pp. 262–63.
30. Ibid., pp. 263–64.
31. Ueda and Tsuchiya, *Machiya—kyōdō kenkyū*, p. 77.
32. Nishiyama, *Nihon no sumai*, p. 59.
33. Tetsuo Tamai, "Edo chōninchi no kenkyū 1-4, *Nihon kenchiku gakkai ronbun hōkokushū*, nos. 252–55.
34. Tokutarō Sakurai, *Minkan shinkō to gendai shakai* (Folk Religions and Contemporary Society), (Tokyo: Hyōronsha, 1971), p. 37. Concerning "estate kami," see Hiroji Naoe, *Yashikigami no kenkyū* (Study of Yashikigami) (Tokyo: Yoshikawa Kōbunkan, 1966).
35. Tenbi Kanai, *Shitsugen saishi—Toyofukihara no shinkō to bunka* (Moor Rituals: Religion and Culture of Toyofukihara) (Tokyo: Hōsei Daigaku Shuppankai, 1975).
36. Hirotarō Ōta, *Shintei Zusetsu Nihon jutakushi* (Revised Illustrated History of Japanese House)(Tokyo: Shōkokusha, 1971), pp.56-60.
37. Yoshinobu Ashihara, *Machinami no bigaku* (The Aesthetics of the Townscape) (Tokyo: Iwanami Shoten, 1979).
38. Ibid., pp. 58–61.
39. Teiji Itō, *Kyoto no dezain—kekkai no bi* (Design in Kyoto: The Beauty of Sacred Boundaries) (Kyoto: Tankōsha, 1966), p. 183.
40. Fumitaka Nishizawa has expressed similar views and discussed at length the manifold meanings of "gaps." See Fumitaka Nishizawa, "'Suki' to kenchiku" (Gaps and Architecture), *Space Modulator 51*, Nippon Sheet Glass Co., 1978.
41. On corner lots at the intersection of two major streets, an establishment sometimes had two storefronts.
42. A formal comparison of two buildings serving such different functions is possible because there is no close correspondence between architectural form and its program. For example, St. Philibert, though it may be closely related to Christianity, could have served just as well as an audience hall for a palace if the altar had been replaced by a throne. What is important here is that both buildings have a hierarchical spatial structure.
43. Hisao Kōyama, "Kindai kenchiku ni okeru tetsu to garasu no kūkan, soshite hikari" (Steel-and-Glass Spaces and Light in Modern Architecture), *Space Modulator 53*, Nippon Sheet Glass Co., 1979.
44. Mircea Eliade, *The Sacred and the Profane: The Nature of Religion*, trans. Willard R. Trask (New York: Harcourt, Brace, 1959).
45. For example, Tom Schumacher, "Contextualism: Urban Ideals and Deformations," *Casabella* 359-360 (1971), pp. 5–6.

46. This does not mean the relationship between layers was uncontrolled. On the contrary, in refined forms of art, sensitivity to a sympathetic relationship between objects or to magnetic-like fields developed. That sensitivity was expressed by the aesthetic concept of *ma*. *Ma* is similar to Western proportions in that it imposes on compositional elements an aesthetic order, but unlike proportions it is not numerically formulated. It is a quality that is "discovered" in nature.

CHAPTER 5

# THE JAPANESE CITY AND INNER SPACE

*Fumihiko Maki*

## INTRODUCTION

To foreign visitors and even to many residents, Tokyo appears to be a paradigm of urban chaos, spatially confusing and structurally illegible. Nevertheless, every city with a long history possesses an internal logic in its physical form, however unclear or inconsistent this logic may appear at first, and Tokyo is no exception. In the chaos of its urban plan, we can read a rich collage of patterns and figures, which relate a colorful, contradictory history of several centuries of urban growth. In preface to a discussion of certain aspects of Japanese space conception and urban form, a brief history of Tokyo may be helpful to those who are unfamiliar with the city.

Tokyo, originally known as Edo, began as a settlement around the tenth-century castle of the feudal lord Ōta Dōkan. The original development took place at the mouth of the Sumida River along Edo Bay, on the fertile Kanto Plain. The city was laid out in accordance with an ideal Chinese model, with Edo Castle at its center. In this model, each cardinal direction carried symbolic meaning and strategic importance that corresponded quite readily to Edo's geographical situation. A spiraling system of canals surrounding Edo Castle served to protect it from attacks but also connected it to the Sumida River to allow easy transport of goods. The flat land between the castle and the river naturally became a district of merchants and craftsmen, known as the *shitamachi* ("low city").

In 1605, Tokugawa Ieyasu selected Edo as the seat of his new shogunate. Direct communication between Edo and the rest of the country become critical, and radial thoroughfares from the castle to outer regions were overlaid on the existing city plan. Important *daimyō* (lords) took up residence on land surrounding Edo Castle, particularly in the hilly *yamanote* ("high city") district to the west. Many large estates occupied hilltops affording commanding views and covered by lush greenery. Shrines and temples also developed in the *yamanote* hills.

By 1710 Edo's population had grown to more than one million, making it at that time the world's largest city. Land ownership was by no means equitable: members of the feudal upper classes, though small in number, lived on nearly three-quarters of the land, while the larger population of craftsmen and merchants occupied a mere fraction of the city. The city plan as a whole was developing into a fairly complex array of overlapping figures and patterns. The flat, densely populated *shitamachi* was characterized by a fairly regular grid plan, intersected by canals and radial roads. Lower-rank samurai also lived in neighborhoods based on grid patterns, though there the streets were spaced further apart. The spacious lands of the *daimyō* were developed more freely and adapted to the irregular topography and natural features of the high city. Edo developed into a spectacular garden city of great scenic variety.

With the Meiji Restoration in 1868, the emperor came to rule in Edo, which was renamed Tokyo ("East Capital"). The new government, intent on Japan's rapid modernization, brought about many changes to the urban form. Railroads, bridges and other civil engineering projects knitted together formerly isolated parts of the city. Following Japan's renewed contact with Western nations, city planners introduced broad avenues and European-style zoning regulations. At first, powerful landowners were able to resist government efforts to develop their land; however, the end of feudal privileges and declining fortunes eventually forced most of them to sell off their property piecemeal. Large estates were divided over time into smaller lots. Although this process of subdivision often introduced a grid pattern, there was generally no attempt to coordinate these separate grids. The resulting city plan was a fragmented patchwork of grids and other patterns.

Destruction wrought by World War II created another opportunity to rebuild the city with a stronger infrastructure; new motorways and rail lines were overlaid on the old plan. At the same time, the old patterns—the spiral, radii, grids, and free forms adapted to topography—remained imprinted on the urban form. The overlapping of plan systems in Tokyo, the fact that no one system can be read as dominant, gives the city an elusive, seemingly chaotic character. Nevertheless, beyond these conflicting patterns, several identifiable spatial and morphological principles have developed continuously over many centuries; familiarity with these principles may enable us to see present-day Tokyo more clearly.

## LAYERS OF SPACE

In sharp contrast with the *yamanote* section of Tokyo, the *shitamachi* Nihonbashi district (where my office was first located) has been a mercantile center ever since the capital was called Edo. The flat, low-lying area extending from Nihonbashi to the Ningyō-chō and Akashi-chō neighborhoods lies on land that was reclaimed piecemeal from the sea. This part of the city has also been razed by fire many times. Neighborhoods are now divided by straight roads into square blocks because of these circumstances. The main streets, lined with medium-rise and high-rise buildings, are today scarcely to be distinguished from main streets in high-city Tokyo. However, as soon as we leave the main streets, we find mostly two-story houses clustered on both sides of narrow roads tucked between taller buildings.

Narrow lanes separate house from house, and in summer reed or bamboo blinds screen overhanging second-floor balconies. Though nothing of the dim interiors can be seen from the street, the faintest hint of movement within makes us aware of layers of space peculiar to Japan. Thus even in the low city, which does not have the topographical variety of high-city areas, we can discover an amazing diversity of spatial forms even today. The terms *omote* (front) and *ura* (back) have traditionally been used in explaining this typically Japanese phenomenon.

There has long been a tendency in Japan, evident in such terms as *omote-dōri* (main street) and *ura-dōri* (back street), *omote-mon* (front gate) and *ura-mon* (back gate), to establish hierarchical relationships that identify dominant and subordinate elements in urban space and living patterns, although such a tendency is by no means limited to this country. However, this seems to me insufficient to explain the essential character of Japanese space, that is, to account for phenomena that are deeply rooted in the collective unconscious of most local communities.

The concept of *omote* and *ura* cannot fully explain the manifold layers of Japanese space. The founding of a city or smaller community is always a matter of collective or individual will. Unlike individual dwellings, a city or community is shaped by norms developed over a long period of time. Of course such norms may be abrogated and different standards established in the face of new external conditions. However, as the history of cities and communities the world over tends to show, norms peculiar to specific areas are few. In fact, the basic patterns of community structure are far smaller in number than the types of buildings arranged in those patterns. Those patterns also change very little in comparison with such external conditions as transportation, social institutions and lifestyles, which undergo constant transformation.

Cities today are undergoing greater changes than ever before. Perplexed by this kaleidoscopic world, we often try to cope by piecemeal treatment of symptoms instead of root causes. We must first understand what is unchangeable or resistant to change in order to reach a true understanding of what we must or can change.

Having traveled to many cities abroad, I am inclined to believe that multilayered spaces are among the few phenomena observable only in Japan. The Japanese have always postulated the existence of what is called *oku* (innermost area) at the core of this high-density space organized into multiple layers like an onion, and the concept of *oku* has enabled them to elaborate and give depth to even a delimited area.

In the formation of urban space, certain stable concepts that have been sifted and committed to memory by the collective unconscious of the community work automatically. *Oku*, a spatial concept peculiar to Japan, is a good example, and I believe an understanding of this way of perceiving space is important in formulating ideas of what future cities should be like.

A centripetal *okusei*, or inwardness, has always been basic to space formation in Japan, from village to metropolis. Understanding the concept of a centrifugal center found in other cultures will make this centripetal inwardness in Japan easier to comprehend. First, however, a brief discussion of *oku*, or inner space, as found in Japanese surroundings is in order.

The word *oku*, expressing a distinctive Japanese sense of space, has long been a part of the vocabulary of daily life, as is evident from such literary classics as the eighth-century poetry anthology the *Man'yōshū* (Collection of Ten Thousand Leaves), the tenth-century *Ise monogatari* (Tales of Ise), the fourteenth-century *Tsurezuregusa* (Essays in Idleness) and *kabuki* plays of the Edo period. It is interesting to note that the use of the term with respect to space is invariably premised on the idea of *okuyuki*, or depth, signifying relative distance or the sense of distance within a given space.

The Japanese, long accustomed to a fairly high population density, must have conceived space as something finite and dense and in consequence developed from early in their history a sensitivity finely attuned to relative distance within a delimited area. Only that can satisfactorily explain the concept of *okuyuki*. The Japanese distinguish an innermost portion even when a space measures only 100 meters—or for that matter only 10 meters—in extent and carefully lay out a route leading to that portion. Only in that context can the idea of multilayered space, and the Japanese attempt to structure space—we might even go so far as to say the Japanese conception of the cosmos—be understood.

*Oku* also has a number of abstract connotations, including profundity and unfathomability, so that the word is used to describe not only physical but psychological depth. It is interesting to note how often the Japanese use the word in adjectival form. Such usages include *oku-dokoro* (inner place), *oku-guchi* (inner entrance), *oku-sha* (inner shrine), *oku-yama* (mountain recesses), and *oku-zashiki* (inner room)—all relevant to the notion of physical space; *oku-gi* (secret or hidden principles) and *oku-den* (secret mysteries of an art), referring to things invisible but present in hidden form; and *o-oku* (wife of a *shōgun*) and *oku-no-in* or *oku-gata* (wife of an aristocrat or nobleman), terms suggesting social position.[1] Evident in the use of all these words is a tendency to recognize and esteem what is hidden, invisible or secret.

The painter and essayist Eiji Usami's *Meiro no oku* (The Inner Labyrinth) is one of the earliest essays on the sense of inwardness in Japanese space. Usami notes this inwardness in the mazelike effect produced by the winding corridors of Japanese-style inns found in tourist resorts and spas, buildings much more spacious inside than their unpretentious exteriors suggest. Usami speculates that this complex division of space not only results from the gradual accretion of extra rooms and extensions or from architectural considerations necessitated by topography and landscape but also conforms to and reflects a Japanese propensity for labyrinths. He writes:

What causes the feeling of weariness and isolation—the exaggerated sense of being far from home—that comes over us when we arrive at an inn and are shown to our room by a maid? Is it a sort of animal sense—the complete submission to the natural continuity of time we feel arising in us as we follow the long corridor, searching with our eyes for landmarks with every twist and turn—that brings our souls to (putting it somewhat floridly) the *état d'âme* of our remote ancestors? Or is it that the changed aspect presented by every turn and the slight irregularity in the rhythm of our footfalls in going up and down stairs gradually lure our minds from reality toward illusion? Does not this sense of distance signify how far we have strayed into a world of fantasy?[2]

## INNER SPACE IN DWELLINGS

Examining inner space in traditional Japanese dwellings after noting its existence in the recesses of mountains and woods, in Japanese-style gardens as artificial and dwarfed transfigurations of nature, and in the narrow backstreets of cities enables us to perceive that *oku* has been given a still more clearly defined location and status in such houses (Fig. 5-1).

Several years back, I had occasion to call on an elderly woman at her modest *machiya* (merchant's townhouse) in Akashi-chō in the Tsukiji district of *shitamachi* Tokyo. Stepping into the house through a latticed door slightly over a meter in width, I found not in front of me but to one side a raised doorsill leading to a small, 4.5-mat (about 7m$^2$) room behind *shōji* panels. In the center of the room was a *kotatsu*, a quilt-covered table built over a sunken

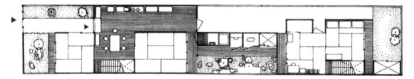

**Fig. 5-1** Plan of a Kyoto *machiya* (townhouse): entering from the street (to the left), one passes in a linear fashion from the most public to the most private rooms of the house, which are set deep into the city block. From Noboru Shimamura, Yukio Suzuka, et al., *Kyō no machiya*

pit. Light entered from the street, which ran parallel to the vestibule through which I had entered. This section of the house was used for receiving guests. Behind this room was a second 4.5-mat room combining the functions of kitchen and bedroom. I was amazed at the complexity of orientation and density of space manifest in a total floor space of only 26 square meters or so. The presence of a household Shinto shrine, a family Buddhist altar, and a *tokonoma* alcove added to the complexity and reinforced the feeling of inwardness and depth.

I was struck by how the orientation of space peculiar to a Japanese-style inn as described by Usami, which leads us ever inward (unlike a Western hotel, which seems to stretch outward from the center) still survives today in modest family dwellings. Moreover, it not only exists as a spatial concept but permeates more abstract social structures by way of the collective unconscious, thus universalizing the concept of inner space.

Why has what might be called a "philosophy of inner space" been cultivated since ancient times in the Japanese cultural tradition?

## THE PROTOTYPE OF INNER SPACE

The Pacific coast of the Japanese archipelago, especially the area stretching from the Kanto plain of central Honshu down to Kyushu, which has been settled since very early times, is favored with a relatively mild climate and abundant water, as well as variegated scenery and forests of evergreen trees. It is popularly believed that these natural features nurtured the typically Japanese outlook on nature. This, of course, refers only to the time since the Yayoi period (c. 200 BC–AD c. 250), when rice cultivation became widespread; before that, our ancestors lived mainly by hunting and gathering and would have had little time for appreciating the beauty of their makeshift mountain abodes.

However, the spread of rice cultivation and the settling of people in the plains in the Yayoi period gave rise to a noteworthy development. A distinction came to be made between lowland villages, where people lived, and mountains, which became special areas, a realm outside ordinary people's sphere of activity. Thus set apart, mountains gradually became more exalted and sacred, until eventually they became taboo areas that were objects of worship. This is how Shinto, an amalgam of animism and shamanism, was born.

As Yūichirō Kōjiro has pointed out in his study of the Japanese community, the prototype of the Japanese rural community is the farming village, comprising rice paddies and a cluster of houses set against a mountain

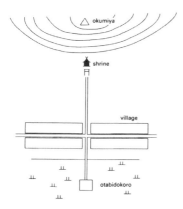

Fig. 5-2 Diagram of an archetypal Japanese village, with its "inner space" hidden deep in the mountain.

backdrop (Fig.5-2).³ This is highly significant, for it graphically suggests the presence of inner space. The typical village has an elongated form that follows a highway skirting mountains and overlooking paddy fields. Perpendicular to this axis is a religious axis, linking the village to a village shrine at the foot of a mountain and an inner shrine (*oku-sha*) in the recesses of the mountain. Here, for the first time, inner space has a religious dimension, in that it suggests the direction in which the seat of a deity (*kami*) lies. This indeed is the prototype for a pattern found throughout the country, including Tokyo: a shrine building, set back slightly from the street, standing in front of a grove of trees.

The inner shrine is located deep in a mountain because it is believed that important things should remain hidden; a winding mountain trail therefore provides the only access. This is in sharp contrast to the European pattern in which the church, the center and symbol of faith, is deliberately built in a conspicuous location (Fig. 5-3). If my characterization of this as an example of inner space is correct, then the concept of *oku* has existed in Japan's local

Fig. 5-3 Panoramic view of a typical rural Japanese village.

communities from remote times. Although in this case inner space may have represented a world set apart, which was an object of worship, it was not entirely divorced from everyday life.

In the seventh century, people began to regard mountain recesses (*oku-yama*) as objects of aesthetic appreciation. Many Japanese mountains appear rounded and gently sloping, especially when viewed from a distance. The idea of incorporating images of mountains and rivers in gardens may not have suggested itself had Japan possessed a different kind of scenery. Would it occur to people in India and Spain, with their rugged, treeless mountains, or in Borneo—an island in the monsoon region like Japan but one located in the tropical zone, with its 4,100-meter Mount Kinabalu, a harsh, blackened peak with a flattened top—to incorporate miniature domains of such landscapes in their gardens? However, the Japanese idea of *oku* is not associated solely with mountains. The word itself is said to be derived from *oki*, meaning offshore waters. Shinobu Origuchi, a scholar of Japanese literature, theorizes that gods were believed to have come from across the sea. If *oku* does indeed imply the seat of a deity, then we can logically conclude that both the mountains and the sea had their respective *oku*, their own inner depths. Had religion in Japan been monotheistic, the idea of *oku* would not have become so pervasive in Japanese life.

By studying the significance of woods and land in Japan's ethnological history, we can get an even better idea of the pervasiveness of the idea of inner space. Many of the mountains regarded as divine are covered with beautiful woods or rise in smooth, rounded shapes. Even wooded hills on plains, regardless of size, are accorded special treatment by local villagers. Ancient burial mounds to be found even in the vicinity of present-day Tokyo were apparently modeled after natural mountains. Often a small shrine is hidden in the trees that cover a mound. Such mounds were probably among Japan's first man-made symbols.

Interestingly, these symbols of power, which the philosopher Takeshi Umehara calls the first monumental structures to be built in Japan, assumed a pseudo-natural appearance. This was true even for land forms constructed on a more modest scale. For example, a small shrine was often erected on a hillock in the garden to house the god of the harvest and other household deities. As this demonstrates, elevated land was considered the seat of the god of the land; foliage represented a secluded natural environment appropriate to a god. Such places suggest, by slight changes in the lay of the land, trees and views, the subtle nuances to be found in nature.

In *Bungaku ni okeru genfūkei* (The Primary Landscape in Japanese Literature), literary critic Takeo Okuno writes that until the early Showa period, folk beliefs were still attached to the small shrines, stone Buddhist images, and stone monuments that were found here and there in open fields in Tokyo's high-city *yamanote* area, where he played as a boy. Okuno writes that, even after folk beliefs such as the god of the land had been absorbed into the beliefs of the larger community in guardian and tutelary gods, certain areas of open fields continued to have a fantastic, magical quality.[4] His observation points out the fact that the Japanese city was (and in all likelihood still is) in certain respects an enormous village and has always been inherently rural in character. Okuno concludes that open fields in the city were taboo areas, not just undeveloped vacant lots. Needless to say, such areas were latent inner spaces.

Japan's Pacific coastal areas generally project a cheerful, benign image. By contrast, Japan's folk history with respect to the land and soil is dark. The interest of the Japanese in the subtlest contours of the land stems from a strong, sometimes preternatunal attachment to the land. People's worship of the land, revealed by myriad stories and legends, has become a part of the collective unconscious. Though the particulars of that worship have been mostly forgotten, attachment to the land remains very much alive even in modern socioeconomic institutions. Thus, any space that functions as a private sanctuary is given ritualized status as an inner space, or *oku*. In no other country have people been so attached to land and so little disposed to regard buildings standing on land as permanent. In Japan, urban space means land, not structures.

## INNER SPACE VERSUS CENTER

A visitor to an old European city, especially a small one, will find the most important and impressive buildings, such as churches and the municipal hall, concentrated in the central district. Unlike a Japanese castle, which stands above its castle town, important buildings in a European city are not isolated but part of the everyday environment, even as they assert their importance. Many churches have soaring spires clearly visible from any part of town. Nowadays most people regard a church spire as nothing more than a landmark, but once it meant much more.

In ancient times the city was a domain of order set apart from chaos. A church spire was an *axis mundi* or cosmic pivot that assured communication with heaven; it was both the center and the symbol of the city. Before there were cities, the cosmic pivot had been represented at various times by trees, mountains, and ladders. In the Islamic tradition the Kaaba of Mecca represents the supreme place on earth; its Christian equivalent is the hill of

Golgotha (Calvary), the site of the Crucifixion. With the introduction of churches and spires, the city itself came to symbolize the center of the world.

Significantly, a center is premised on the existence of a uniform space around it. Inner spaces do exist in European cities. However, the moment the idea of uniformity of space is introduced, the concept of inner space in the Japanese sense is no longer tenable and gives way to the more universal and easily comprehended concept of a center. A center, unlike inner space, must be open and visible, like the hilltop church spire that rises toward heaven. The concept of center is not limited to the West and its Judeo-Christian tradition but is also found in China and South Asia.

The Japanese too worshiped mountains and regarded trees as sacred. When and why did the Japanese part ways with the rest of the world? First, the ancestors of the Japanese did not regard mountains as absolutes. They revered deep mountain recesses but also accorded respect to nearby mountains and hills. They built towns and villages in valleys tucked between mountains or in basins ringed by mountains, perhaps out of a belief that those surrounding mountains were themselves guardian gods.

The Japanese identified as the point of origin, not the summit of a mountain, but the depths of mountains. In the West, the idea of a center (such as a mountaintop) was expressed in the form of cities, churches, and spires, whereas in Japan mountains were expressed—in tumuli and gardens—in such a way as to suggest inner space, not a cosmic pivot (Fig. 5-4).

The argument may be made that the Japanese shunned thickly wooded mountains as sites for settlements because of the nature of the evergreen forest zone in which they lived, or that they chose to inhabit valleys because of easy access to water. However, I believe such utilitarian considerations served merely to reinforce the Japanese concept of space.

A center locates a vertical, cosmic pivot that directly links heaven and earth. A culture with towers is premised on that idea. In the West, verticality

Fig. 5-4 Approach to the town of Assisi, Italy.

enhances the majesty of churches. Standing in front of a Gothic cathedral we are overwhelmed by its almost superhuman vertical scale. Stepping into its dim interior through low, heavy doors, we find ourselves in a soaring space; slanting rays of sunlight filtered through tall stained-glass windows suggest infinite vertical extension. The drama is intended to be both physically and spiritually exhilarating. Renaissance churches, as a result in part of a Hellenistic influence, are more human in scale. Nonetheless, they too were characterized by centrality and verticality. Dancing angels painted in fresco under the central dome of a Renaissance church create, in the dimness of light, an illusion of spiral upward movement.

Some years ago a friend of mine drove me to the ancient city of Urbino in central Italy. As we were approaching the road leading to the heart of the city, my friend suddenly pointed toward a hilltop across the valley. I saw a church directly lit by the rays of the setting sun; its entrance arch was bathed in gold, creating a striking contrast with its white walls. For those who designed the church, even the orientation seems to have played a part in its dramatic symbolism.

Inner space emphasizes horizontality and gains symbolic power by concealment. A Shinto shrine is, therefore, not a space to be entered but an object to be seen from without. The ridgepole of the shrine symbolizes a sacred tree, the open veranda circling the shrine in all likelihood the fence that once surrounded a sacred pillar. The shrine stands silently, wrapped in trees. If the location is deep in the mountains, mist can gather at times and obscure even the sight of the shrine, drawing us into a world of ephemerality and flux.

The absence of centrality and verticality in Japanese architecture is especially evident in the treatment of pagodas. The cultures of the world can be divided into those with towers and those without towers. Japan probably belongs to the towerless cultural sphere. Pagodas were introduced into Japan

Fig. 5-5 *Yabu Lane below Mt. Atago*, woodblock print. From Hiroshige Ando's *Hundred Views of Edo*.

from China in mid-sixth century as symbols of Buddhist culture.

The philosopher Takeshi Umehara mentions a strange tower he saw in Xi'an (the ancient capital of Ch'ang-an) on a visit to China some years ago. He comments that the sight of this red-brick tower—built by the Buddhist priest and traveler Xuanzang (602–664)—soaring into the blue heavens must have awed people of the time.[5] As his account indicates, in their centrality and verticality, spaces in Chinese civilization more closely resemble the spaces of western Asia and Europe than they do those of Japan, though Japan and China are both in East Asia.

As can be seen at Hōryūji, when the pagoda was introduced into Japan, it lost its upward impulse and became simply one of many elements—such as the main hall, roofed galleries and even trees—used to achieve overall equilibrium. As Umehara points out, the Hōryūji pagoda resembles a standing human figure. The pagoda is nothing more than a vertical accent in an ensemble characterized by a Japanese sense of balance.

Inner space is a mental touchstone for those who observe or produce it. In that sense inner space can be called an invisible center, or more precisely, a convenient alternative to the center, devised by a culture that denies absolutes such as centers. People are free to decide for themselves what constitutes such a "center"; there is no need to make it explicit. The multilayered structuring of space, one of the compositional patterns distinctive to Japan, gradually developed in this way.

As an ultimate destination, innermost space often lacks a climactic quality. Instead, it is the process of reaching that goal that demands drama and ritual. The design of an approach is a matter of manipulating horizontal depth rather than height. The approaches to many temples and shrines turn and twist, with trees and slight undulations in the ground, now revealing, now concealing the goal. This structuring of spatial experience takes into account the dimension of time. Even the *torii* gate at the entrance to a Shinto shrine is an element in this ritual of arrival.

Nowadays most shrines and temples in big cities have sold their lands and lost every trace of their *sandō*—the road that led to the main temple or shrine building—and the grove that stood in the back of the compound or precinct; only the buildings themselves remain, their inner recesses laid bare. Does inner space inevitably become empty when exposed? Is it scattered and made null? Japanese culture has no historical experience of ruins in the true sense of the word. Any temple or shrine building could be restored, at times with great ease, if destroyed, and was therefore not meant to be permanent.

This reminds me of something I saw in the American city of St. Louis many years ago. The city had developed as a trading post on the west bank of the Mississippi River around the end of the eighteenth century. By the mid-twentieth century, the heart of the city had become a slum inhabited mostly by minorities. When I was there, an extensive slum clearance project covering dozens of blocks was in progress as part of an ambitious city redevelopment plan. A project of this kind usually entails moving the church buildings, as well as the residents, out of the affected district, though Catholic churches, unlike Protestant churches, tend to opt to stay on. In that desolation of razed lots a single Catholic church remained standing, completely isolated but in full possession of a dignity that had nothing to do with the building's size or design. I still remember vividly the robustness of that church, which gave an impression so different from that of a naked Japanese shrine.

## ENVELOPMENT VERSUS DEMARCATION

The difference between inner space and a center will be made still clearer by a comparison of concepts of place and formation of territory. The 1954 film *Le Grand Jeu*, shown in Japan many years ago, depicts life in an oasis set in the bleak, empty desert beneath a scorching sun. The oasis, with its meager greenery and scant water, its fleeting freedom and romance, is a contemporary desert paradise girded by walls and gates of sun-baked brick and stone.

As previously discussed, sanctification of place, determination of a center as cosmic pivot, and establishment of a cosmos within chaos are the archetypal acts by which cities have been formed in much of the world. The deserts in which nomads wandered and the Aegean Sea where the ancient Greeks sailed were always filled with uncertainty and danger. To make the city a haven and a paradise, it was necessary not only to determine its center but also to demarcate clearly its surrounding territory.

The ancient Greeks when founding a city performed a ceremony fixing the city limits; to them the city walls themselves had a sacred significance. In a civilization in which the formation of a city was premised on the existence of a center, the building of encompassing walls was a necessary act. It is also important to note that a rational spirit informed the patterns used to subdivide the territory within a city.

We can clearly see the difference in pattern between the grid-type city of Miletus, believed to have been built by Hippodamus in the fifth century BC, and a cluster-type mountain city of medieval Europe. The former reveals itself by its very form to be a city-state inhabited by free citizens, whereas the layout of the latter expresses the hierarchical order of an ecclesiastical, class-oriented society. Nonetheless, each had a clearly fixed center and boundaries and was established as a finite entity in a limitless expanse of space. Furthermore, the

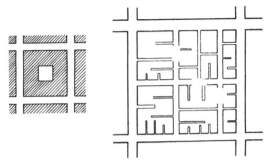

Fig. 5-6 Typical street and block divisions in a *shitamachi* area of Tokyo. From Uzo Nishiyama, *Nihon no sumai*.

demarcated territory was well organized from an overall viewpoint. Thus the two types of cities, despite differences in sociopolitical background and physical layout, have basically corresponding structures.

By contrast, city planning in Japan was based on an entirely different approach. As already mentioned, the Japanese, in undertaking the building of a city, invariably recognized the finite nature of land. Thus many cities were founded in basins ringed by mountains despite their geographical disadvantages for defensive purposes; for the same reason, it was exceedingly rare for a Japanese city to have man-made boundaries like those that defined so many cities in other parts of the world. In Japan, instead of a fixed center, territorial integrity was based on something indeterminate, and enveloping or enfolding this basic "something" (*oku*) was the operational principle of territorial formation. In contrast to active demarcation, enveloping implies passivity as well as flexibility—that is, a capacity to adapt the envelope to the form of what is to be enveloped.

Several years ago I had the opportunity to see an exhibition of traditional Japanese containers and wrapping materials held in New York under the auspices of the Japan Society. I was struck anew by the profusion of traditional wrapping materials and methods. There were containers of cloth, straw, paper, leaves and other materials. I know of no other civilization that has developed a system of wrapping so beautifully and functionally adapted to the nature and shape of the objects to be wrapped.

It seems to me that the principle of "inner space envelopment" in the formation of territory is a major Japanese concept, corresponding to that of "center demarcation" in other cultures. It is true that a grid pattern was generally used in laying out flatland portions of cities during the Edo period. But the Japanese version lacked the internal consistency of a grid like that of Miletus, which maintained a single system of coordinates even though the

Fig. 5-7 Pattern of roads in an actual *yamanote* neighborhood of Tokyo, shown according to their chronological development: heavy lines indicate roads enveloping the area (originally a single property); thinner black lines show main roads leading in from the boundaries, and dotted lines are minor roads added later.

city was divided into two parts by an arm of the sea. Japanese cities, on the contrary, were highly susceptible to the influence of topographical changes; as a result, geometrical conformity to a grid was not easily sustained.

Fujimi-chō, a district so named for its fine view of Mount Fuji, is a case in point. Because the district's roads, forming a grid, are purposely oriented in relation to Mount Fuji, they are not aligned with the roads of the adjacent districts. In other words, instead of a portion of the theoretically infinite expanse of the grid being encapsulated, a certain area with a common characteristic—in this case, a view of Mount Fuji—is identified as a territory by "wrapping it up" as a separate grid of several blocks.

It goes without saying that in Edo, where layout changed from place to place even in the flat downtown districts (Fig. 5-6), the formation of territory by enfolding or enveloping it was still more pronounced in the high-city *yamanote* districts (Fig. 5-7). There the roads running along hilltops and through valleys provided the basic lines of territorial delineation, clearly functioning to envelop the area following its natural contours. The narrow lanes leading from these outer "border" roads to the interior of the territory thus enclosed did not intersect. In this way disturbance of the integrity of the territory as a whole was minimized, and the "inner space" of such areas has remained intact to date.

The winding roads of medieval European castle towns, though seemingly analogous, obey an essentially different set of structural imperatives. After all, whether laid out as grids or as networks, Japanese cities are intrinsically Japanese, and Western cities, Western. The contrasting concepts of "center demarcation" and "inner space envelopment" show the basic difference in the way space is organized in the two types of civilizations.

Using the idea of inner space as a clue, we have now finally come to the essence of the quality of "place" in two types of civilizations. As shown by the stratagems employed in forming territories in space (be it the desert, wilderness or ocean)—that is, cosmic pivot, center, enclosure and demarcation—people belonging to center-oriented civilizations believe that only what is made can exist absolutely, space itself being inherently formless and infinite.

I have already pointed out that Japanese cities are like villages in that they incorporate rural institutions and landscapes. I believe that Japanese cities have grown out of the soil instead of being made by carving a measure of abstract space and architecture out of an infinite expanse of space, as is the case with cities in center-oriented civilizations.

For the Japanese, land is a living entity; that is the basis of their reverence for land, a feeling deeply rooted in folk beliefs. Surely inner space is not something constructed, like a center, but something bestowed by the land itself. The Japanese do not hesitate to demolish houses, perhaps because a house is, after all, no more than a temporary abode in a transient world. But they are averse to the removal of wells or mounds.

Once the concept of inner space was universalized, inner space (*oku*) in houses became nothing more than a specified, relative location within interior space. Surely this was because the Japanese saw in land the source of existence, and inner space was only its symbol or proxy. Thus towns and villages generate countless inner spaces. The city can be seen as an aggregate of innumerable public and private territories, each enveloping its own inner space. The Japanese city developed not as a community clustered around an absolute center but as numerous territories, each safeguarding its own inner space—be it public, semi-public or private. Japanese cities maintained this form of organization at least until the early decades of the twentieth century.

Today Japanese cities are being subjected to unprecedented modernization and growing population density. The purpose of this essay is not to discuss the past development or predict the future of inner space in Japan. As the French sociologist Maurice Halbwachs has remarked, a city is a place of collective memory.[6] My aim here is to point out how indispensable a knowledge of the cultural images rooted in the collective unconscious of the community is to an understanding of the nature of the city.

An architect like myself, who plays a role, however limited, in the building of modern cities, is faced with an inescapable question. Various basic scenarios for cities—including scenarios for hell—can be easily imagined. The question is how those scenarios interact with reality. According to one scenario, ever-increasing urban population density leads to the further loss of an already stunted nature and of the sense of place rooted in land, resulting in the dispersion of inner space (including the "exposed inner space" discussed earlier). Inner space becomes more and more compartmentalized, being relegated to one portion of an apartment, for instance, and thus ceases to participate in the kind of collective inner space formerly found in both low-city and high-city sections of Tokyo.

In another scenario, efforts are made to revive urban inner spaces wherever possible, utilizing all available spatial concepts and techniques, old and new. What form such revived inner spaces should take is still uncertain. But I am convinced that once the goal is defined, we will discover the means to attain it. The history of Japanese cities teaches us that the qualities desired in space are to be achieved through, not just expansion, but also the creation of depth.

NOTES

1. The last three terms are derived from the wife's quarters being in the inner part of the residence.
2. Eiji Usami, *Meiro no oku* (The Inner Labyrinth) (Tokyo: Misuzu Shobō, 1975), pp. 211–12.
3. Yūichirō Kōjiro, *Nihon no komyūnitii* (The Japanese Community), Space Design no. 7 (November 1975), pp. 8–12.
3. Takeo Okuno, *Bungaku ni okeru genfūkei* (The Primary Landscape in Japanese Literature) (Tokyo: Shūeisha, 1972), pp. 85–86.
4. Takeshi Umehara's description of the awe-inspiring tower in Xi'an is recalled from a lecture.
5. Maurice Halbwachs, *La mémoire collective* (Paris: Presses Universitaires de France, 1950).

CHAPTER 5:

First published in *Nurturing Dreams: Collected Essays on Architecture and the City* by Fumihiko Maki, edited by Mark Mulligan, published by The MIT Press, 2008.

# Figure source

## CHAPTER 1

**Fig. 1-1** MIT Rotch Library.
**Fig. 1-2** Kevin Lynch, *The Image of the City*, Harvard-MIT Joint Center for Urban Studies Series (Cambridge: The MIT Press, 1960).
**Fig. 1-3** Rem Koolhaas, *Delirious New York* (New York: Oxford Press, 1978), p. 243.
**Fig. 1-4** Stanford Anderson, "Studies toward an Ecological Model of the Urban Environment," in Stanford Anderson, ed., *On Streets* (Cambridge: The MIT Press, 1978), p. 301.
**Fig. 1-5** Regional Plan Association, *Urban Design Manhattan* (London: Studio Vista, 1965), p. 61.
**Fig. 1-6** Toshi dezain kenkyūtai, *Nihon no toshi kūkan* (Urban Spaces of Japan) (Tokyo: Shōkokusha, 1965).

## CHAPTER 2

**Fig. 2-1** Akira Naitō, *Edo to Edojō*, (Tokyo: Kajima Institute Publishing Co., Ltd, 1966).
**Fig. 2-2~6, 8** Based on Yasuo Masai, *Edo no toshiteki tochiriyōzu 1860nen goro* (Urban Land Use Map Ca. 1860), 1973.
**Fig. 2-7** *Owariyaban Kaei Keiō Edo kiriezu*, in Mikito Ujiie and Takashi Hattori, *Yamakawa MOOK Edo Tokyo kiriezu sanpo* (*Yamakawa MOOK Edo Tokyo kiriezu stroll*) (Tokyo: Yamakawa Shuppansha, 2010).
**Fig. 2-9** *Bushū Toshimagun Edo shōzu*, 1632.
**Fig. 2-11** Uzō Nishiyama, *Nihon no sumai* (Dwellings of Japan) (Tokyo: Keisō Shobō, 1976).
**Fig. 2-16** *Owariyaban Kaei Keiō Edo kiriezu, 21, Azabu ezu, Yamakawa MOOK Edo Tokyo kiriezu sanpo*.
**Fig. 2-18** Based on Minatoku Mita Toshokan, *Minatoku henkaku zushū*.
**Fig. 2-19** *Owariyaban Kaei Keiō Edo kiriezu, 26, Otowa ezu, Yamakawa MOOK Edo Tokyo kiriezu sanpo*.
**Fig. 2-20** *Naimushō Chirikyoku chikeizu*.
**Fig. 2-21** *Dainihon Teikoku Rikuchi Sokuryōbu ichimanbunnnoichi chikeizu*, Waseda.
**Fig. 2-27** Akira Naitō, *Edo to Edojō*.
**Fig. 2-26** Sigfried Giedion, *Space, Time and Architecture* (Cambridge: Harvard University Press, 1941).
**Fig. 2-28** Top: *Owariyaban Kaei Keiō Edo kiriezu, 7 Kyobashi Minami Tsukiji ezu, Yamakawa MOOK Edo Tokyo kiriezu sanpo*. Bottom: *Owariyaban Kaei Keiō Edo kiriezu, 18 Honjo ezu, Yamakawa MOOK Edo Tokyo kiriezu sanpo*.
**Fig. 2-36** Hisashi Sasaki, *Kenchiku no gunzōkeiron no tameno ichikōsatsu—kaijō shūraku no bunseki* (A Study into Group Form in Architecture: An Analysis of a Cluster-Shaped Village), master's thesis, Faculty of Engineering, University of Tokyo.

## CHAPTER 3

**Fig. 3-1** Suzuki Masao, *Edo no kawa—Tōkyō no kawa* (The Rivers of Edo, The Rivers of Tokyo) (Tokyo: Nihon Hōsō Shuppan Kyōkai, 1978).
**Fig. 3-2** Tadahiko Higuchi, *Keikan no kōzō*: the structure and compositional elements of each of the seven types.
**Fig. 3-3** *Owariyaban Kaei Keio Edo kiriezu, Azabu Hongo ezu, Yamakawa Mook Edo Tokyo kiriezu sanpo*
**Fig. 3-4** Edo *kiriezu* of 1861.
**Fig. 3-5** *Owariyaban Kaei Keio Edo kiriezu, 14 Hongo Yushima ezu* (Jinbunsha)
**Fig. 3-6** Nihon Chishi Kenkyujo, ed., *Nihon Chishi 7* (Topography in Japan, Vol. 7) (Tokyo: Ninomiya Shoten, 1967).
**Fig. 3-11, 12** *Owariyaban Kaei Keio Edo kiriezu, 9 Akasaka ezu* (Jinbunsha)
**Fig. 3-18** *Edo meisho zue*.
**Fig. 3-20** Edmund N. Bacon, *Design of Cities* (London: Thames & Hudson, 1967).
**Fig. 3-21** (The map of park in Tokyo) (Tokyo: Kajima Institute Publishing Co., Ltd., 1974).
**Fig. 3-22** Yūichirō Kōjiro, *Nihon no komyūnitii* (Communities of Japan) Space Design, no. 7 (Tokyo: Kajima Institute Publishing Co., Ltd., 1975).
**Fig. 3-23** Atsushi Ueda, *Ningen no tochi* (Land of Human Beings) (Tokyo: Kajima Institute Publishing Co., Ltd, 1974).
**Fig. 3-24** Yūichirō Kōjiro, *Nihon no komyūnitii*. (Tokyo: Kajima Institute Publishing Co., Ltd., 1975).

## CHAPTER 4

**Fig. 4-16** Noboru Shimamura, Yukio Suzuka, et al., *Kyō no machiya* (The *Machiya* of Kyoto) (Tokyo: Kajima Institute Publishing Co., Ltd., 1971), p. 87.
**Fig. 4-18** *Edo meisho zue*.
**Fig. 4-20** *Gōkan ehankiri kakushino fumitsuki*.
**Fig. 4-21** Akira Naitō, *Edo to Edojō* (Edo and Edo Castle) (Tokyo: Kajima Institute Publishing Co., Ltd, 1966),

p. 162.
**Fig. 4-22** Kiyoshi Hirai, *Nihon no kinsei jūtaku* (Feudal Residences of Japan) (Tokyo: Kajima Institute Publishing Co., Ltd, 1968).
**Fig. 4-23** Naomichi Ishige, *Jūkyo kūkan no jinruigaku* (Anthropology of Residential Space) (Tokyo: Kajima Institute Publishing Co., Ltd, 1971), p. 259.
**Fig. 4-24** Serge Chermayeff, Christopher Alexander, *Community and Privacy* (New York: Doubleday, 1963), p. 217.
**Fig. 4-25** Atsushi Ueda, Atsuo Tsuchiya, ed., *Machiya—kyōdo kenkyū* (Machiya: Collaborative Studies) (Tokyo: Kajima Institute Publishing Co., Ltd, 1975), p 214.
**Fig. 4-26** Ueda, Tsuchiya, *Machiya—kyōdo kenkyū*, p. 204.
**Fig. 4-27** Shimamura, Suzuka, *Kyō no machiya*, p. 177.
**Fig. 4-28** Makoto Takeuchi, "Kansei-Kaseiki Edo ni okeru shokaisō no dōkō" (Class Trends in Edo from the Kansei era to the Bunka, Bunsei eras) in Shonosuke Nishikawa, ed., *Edo chōnin no kenkyū daiikkan* (Studies of Edo Chonin, Vol. 1) (Tokyo: Yoshikawa Kobunkan, 1972), p. 396.
**Fig. 4-29** Nihon Kenchiku Gakkai, ed., *Nihon Kenchikushizushu, kaiteishinpan* (Collected Drawings from Japanese Architectural History, New and Revised Edition) (Tokyo: Shōkokusha, 1963), p. 70.
**Fig. 4-30** Yukio Suzuka, *Kya no machiya*
**Fig. 4-35** Edward S. Morse, *Japanese Homes and Their Surroundings* (Boston: Ticknor and Company, 1886), p. 54.
**Fig. 4-46** Yoshinobu Ashihara, *Machinami no bigaku* (The Aesthetic of the Streetscape) (Tokyo: Iwanami Shoten, 1979), p. 60.
**Fig. 4-53** Peter Murray, *Renaissance Architecture* (New York: Abrams, 1971), p. 172.
**Fig. 4-57** K.J. Conant, *Carolingian and Romanesque Architecture 800 to 1200* (London: Penguin Books, 1959).
**Fig. 4-63** Christian Norberg-Schulz, *Late Baroque and Rococo Architecture* (New York: Abrams, 1974), p.59.

CHAPTER 5

**Fig. 5-1** Noboru Shimamura, Yukio Suzuka, et. al., *Kyō no machiya* (*Machiya* of Kyoto) (Tokyo: Kajima Institute Publishing Co., Ltd, 1991).
**Fig. 5-5** Hiroshige Ando's *Hundred Views of Edo* series.
**Fig. 5-6** Uzo Nishiyama, *Nihon no sumai* (Tokyo: Keiso Shobo, 1976).

# Profiles

Fumihiko Maki / architect
1928  Born in Tokyo
1952  Bachelor of Architecture, University of Tokyo
1954  Master of Architecture, Graduate School of Design, Harvard University
1956–65  Taught at Washington University, St. Louis and GSD Harvard University
1979–89  Professor, Department of Architecture, University of Tokyo
1993  The Pritzker Architecture Prize
1999  Premium Imperiale, The Japan Art Association

1980–99  *Contemporary Japanese Architects : Fumihiko Maki 1-4*, Kajima Institute Publishing Co., Ltd
1980  *Miegakure suru toshi*, Kajima Institute Publishing Co., Ltd (co-authored, Japanese)
2008  *Nurturing Dreams*, The MIT Press


Yukitoshi Wakatsuki / architect
1947  Born in Tokyo
1971  Bachelor of Architecture, University of Tokyo
1973  Master of Architecture, University of Tokyo
1973–present  Maki and Associates
1990  Named Senior Associate, Maki and Associates
1982–92  Visiting Critic, School of Architecture, Tokyo University of Science
1993–97  Visiting Critic, School of Environmental Information, Keio University

1973  *Form of Housing, Mediterranean Region,* Kajima Institute Publishing Co., Ltd (co-authored)
1975  *The Dwelling,* Kajima Institute Publishing Co., Ltd (translator)
1980  *Miegakure suru toshi*, Kajima Institute Publishing Co., Ltd (co-authored, Japanese)
1981  *The Architect: Chapters in the History of the Profession*, Nikkei McGraw-Hill (co-translator)


Hidetoshi Ohno / architect
1949  Born in Gifu
1972  Bachelor of Architecture, University of Tokyo
1975  Master of Architecture, University of Tokyo
1997  Doctor of Engineering (Architecture), Graduate School of Engineering, University of Tokyo
1983–88  Assistant  Professor, University of Tokyo
1988–99  Associate Professor (Architectural Design and Urban Design), University of Tokyo
1998  Research Fellow, Technical University of Delft, Netherlands
1999–2015  Professor (Environmental Studies, Architectural Design and Urban Design), University of Tokyo.
2011  The Prize of AIJ 2011
2015  Emeritus Professor, University of Tokyo

1980  *Miegakure suru toshi*, Kajima Institute Publishing Co., Ltd. (co-authored, Japanese)
1992  *Hong Kong: Alternative Metropolis*, SD9203, Kajima Institute Publishing Co., Ltd. (Japanese / English)
2012  *Design Innovation of Practice*, pp. 67–90, Routledge, London, Print
2016  *Fibercity: A Vision for Cities in the Age of Shrinkage*, University of Tokyo Press


Tokihiko Takatani / architect
1952  Born in Kagawa
1976  Bachelor of Urban Design, University of Tokyo
1976–89  Maki and Associates
1989  Founded Tokihiko Takatani and Associates
2005–  Professor, Graduate School of Tohoku Koeki University
2006  Public Building Award for Makuhari Bay Town Community Center
2011  31st Award of AIJ Tohoku for Fujisawa Shuhei Memorial Museum
2013  The AIJ Annual Design Commendation for Tsuruoka Town Cinema

1980  *Miegakure suru toshi*, Kajima Institute Publishing Co., Ltd. (co-authored, Japanese)
1987  "La genese d'un Labyrinthe" in *Cahiers du Japon, Tokyo au passé et au présent. Japan Echo* (French)

2011  *Revitalization of Small Cities by Social Enterprises*, Gyosei Publishing, Tokyo (ed. Japanese)
2012  *Urban Design Center: A Platform Open to the Public*, Riko Publishing, Tokyo (co-authored, Japanese)

Naomi Pollock / architect, writer
1959  Born in Chicago, Illinois
1981  Bachelor of Arts, Dartmouth College
1985  Master of Architecture, Harvard Graduate School of Design
1990  Master of Architecture, University of Tokyo

1982–83  Cass & Pinnell Architecture, Washington DC
1985–88  Davis Brody & Associates, New York

1998  *Japan 2000: Architecture and Design for the Japanese Public*, Prestel (co-authored)
2005  *Modern Japanese House*, Phaidon Press
2009  *Hitoshi Abe*, Phaidon Press
2012  *New Architecture in Japan*, Merrell Publishers (co-authored)
2012  *Made in Japan: 100 New Products*, Merrell Publishers
2015  *Jutaku: Japanese Houses*, Phaidon Press
2016  *Sou Fujimoto*, Phaidon Press
1990–present  Tokyo Correspondent, *Architectural Record*

Hiroshi Watanabe / writer, translator
1944  Born in Tokyo
1966  Bachelor of Architecture, Princeton University
1971  Master of Architecture, Yale University
1971–77  Maki and Associates
1981–95  Tokyo Correspondent, *Progressive Architecture*

1985  *Space in Japanese Architecture*, Weatherhill (translator)
1991  *Amazing Architecture from Japan*, Weatherhill
2000  *Waro Kishi: Buildings and Projects*, Axel Menges
2001  *The Architecture of Tokyo*, Axel Menges
2002  *The Construction and Culture of Architecture Today*, Ichigaya Shuppan (translator)
2007  *Anti-Object*, Aa Words (translator)
2008  *Nurturing Dreams*, The MIT Press (co-translator)

本書は当社刊SD選書「見えがくれする都市」を加筆・訂正し英語版として刊行するものです。

## City with a Hidden Past

2018年 3月15日　第一刷発行
2019年12月25日　第二刷

| | |
|---|---|
| 著者 | 槇 文彦+若月幸敏+大野秀敏+高谷時彦+ナオミ・ポロック |
| 英訳者 | 渡辺 洋 |
| 発行者 | 坪内文生 |
| 発行所 | 鹿島出版会 |
| | 〒104-0028 東京都中央区八重洲2丁目5番14号 |
| | 電話 03-6202-5200　振替 00160-2-180883 |
| デザイン | 高木達樹 |
| 印刷・製本 | 三美印刷 |

© Fumihiko Maki, Yukitoshi Wakatsuki, Hidetoshi Ohno, Tokihiko Takatani, Naomi Pollock, 2018
ISBN978-4-306-04661-0 C3052
Printed in Japan

落丁・乱丁本はお取替えいたします。
本書の無断複製（コピー）は著作権法上での例外を除き禁じられております。
また、代行業者などに依頼してスキャンやデジタル化することは、たとえ個人や家庭内の利用を目的とする場合でも著作権法違反です。

URL:http://www.kajima-publishing.co.jp
E-mail:info@kajima-publishing.co.jp